Holt Mathematics

Chapter 7 Resource Book

HOLT, RINEHART AND WINSTON

A Harcourt Education Company

Orlando • Austin • New York • San Diego • London

Copyright © by Holt, Rinehart and Winston

All rights reserved. No part of this publication may be reproduced or transmitted in any form or by any means, electronic or mechanical, including photocopy, recording, or any information storage and retrieval system, without permission in writing from the publisher.

Teachers using HOLT MATHEMATICS may photocopy complete pages in sufficient quantities for classroom use only and not for resale.

Printed in the United States of America

If you have received these materials as examination copies free of charge, Holt, Rinehart and Winston retains title to the materials and they may not be resold. Resale of examination copies is strictly prohibited and is illegal.

Possession of this publication in print format does not entitle users to convert this publication, or any portion of it, into electronic format.

ISBN 0-03-078223-6

7 8 9   170   09

CONTENTS

Blackline Masters

Parent Letter	1
Lesson 7-1 Practice A, B, C	3
Lesson 7-1 Reteach	6
Lesson 7-1 Challenge	8
Lesson 7-1 Problem Solving	9
Lesson 7-1 Reading Strategies	10
Lesson 7-1 Puzzles, Twisters & Teasers	11
Lesson 7-2 Practice A, B, C	12
Lesson 7-2 Reteach	15
Lesson 7-2 Challenge	16
Lesson 7-2 Problem Solving	17
Lesson 7-2 Reading Strategies	18
Lesson 7-2 Puzzles, Twisters & Teasers	19
Lesson 7-3 Practice A, B, C	20
Lesson 7-3 Reteach	23
Lesson 7-3 Challenge	24
Lesson 7-3 Problem Solving	25
Lesson 7-3 Reading Strategies	26
Lesson 7-3 Puzzles, Twisters & Teasers	27
Lesson 7-4 Practice A, B, C	28
Lesson 7-4 Reteach	31
Lesson 7-4 Challenge	32
Lesson 7-4 Problem Solving	33
Lesson 7-4 Reading Strategies	34
Lesson 7-4 Puzzles, Twisters, & Teasers	35
Lesson 7-5 Practice A, B, C	36
Lesson 7-5 Reteach	39
Lesson 7-5 Challenge	40
Lesson 7-5 Problem Solving	41
Lesson 7-5 Reading Strategies	42
Lesson 7-5 Puzzles, Twisters & Teasers	43
Lesson 7-6 Practice A, B, C	44
Lesson 7-6 Reteach	47
Lesson 7-6 Challenge	48
Lesson 7-6 Problem Solving	49
Lesson 7-6 Reading Strategies	50
Lesson 7-6 Puzzles, Twisters & Teasers	51
Lesson 7-7 Practice A, B, C	52
Lesson 7-7 Reteach	55
Lesson 7-7 Challenge	56
Lesson 7-7 Problem Solving	57
Lesson 7-7 Reading Strategies	58
Lesson 7-7 Puzzles, Twisters & Teasers	59
Lesson 7-8 Practice A, B, C	60
Lesson 7-8 Reteach	63
Lesson 7-8 Challenge	64
Lesson 7-8 Problem Solving	65
Lesson 7-8 Reading Strategies	66
Lesson 7-8 Puzzles, Twisters & Teasers	67
Lesson 7-9 Practice A, B, C	68
Lesson 7-9 Reteach	71
Lesson 7-9 Challenge	72
Lesson 7-9 Problem Solving	73
Lesson 7-9 Reading Strategies	74
Lesson 7-9 Puzzles, Twisters & Teasers	75
Lesson 7-10 Practice A, B, C	76
Lesson 7-10 Reteach	79
Lesson 7-10 Challenge	80
Lesson 7-10 Problem Solving	81
Lesson 7-10 Reading Strategies	82
Lesson 7-10 Puzzles, Twisters & Teasers	83
Answers to Blackline Masters	84

Date _____

Dear Family,

In this chapter, your child will learn about ratios, proportions, and percents. This includes learning to work with scale drawings, converting decimals and fractions to percents, using percents to solve problems, calculating simple interest, and using scale when reading a map.

Ratios are used to compare a part to a part, a part to a whole, or the whole to a part.

Write three equivalent ratios to compare the number of stars with the number of moons in the pattern.

$\dfrac{\text{number of stars}}{\text{number of moons}} = \dfrac{4}{6}$ *There are 4 stars and 6 moons.*

$\dfrac{4}{6} = \dfrac{4 \div 2}{6 \div 2} = \dfrac{2}{3}$ *There are 2 stars for every 3 moons.*

$\dfrac{4}{6} = \dfrac{4 \cdot 2}{6 \cdot 2} = \dfrac{8}{12}$ *If you double the pattern, there will be 8 stars and 12 moons.*

So, $\dfrac{4}{6}$, $\dfrac{2}{3}$, and $\dfrac{8}{12}$ are equivalent ratios.

Use a table to find three ratios equivalent to 4 to 7.

Original ratio	4·2	4·3	4·4
4	8	12	16
7	14	21	28
	7·2	7·3	7·4

Multiply the numerator and the denominator by 2, 3, and 4.

So, the ratios 8 to 14, 12 to 21, and 16 to 28 are equivalent to 4 to 7.

Your child will learn to write and solve proportions. A **proportion** is an equation that shows two equivalent ratios.

Find the missing value in the proportion.

$\dfrac{3}{4} = \dfrac{n}{16}$

$\dfrac{3}{4} \diagup \dfrac{n}{16}$ *Find the cross products.*

$4 \cdot n = 3 \cdot 16$ *The cross products are equal.*

$\dfrac{4n}{4} = \dfrac{48}{4}$ *Divide both sides by 4 to undo the multiplication.*

$n = 12$

Holt Mathematics

Your child will learn to read and use scales on maps and scale drawings. A **scale drawing** is a drawing of a real object that is proportionally smaller or larger than the real object.

The scale on a map of Yosemite National Park is 2 in. : 1 mi. This means that 2 inches on the map represent 1 mile in the park. On the map, the distance between North Pines Campground and Nevada Falls is 4 in. What is the actual distance?

$\frac{2 \text{ in.}}{1 \text{ mi}} = \frac{4 \text{ in.}}{x \text{ mi}}$ Write a proportion using the scale. Let x be the actual number of miles from North Pines Campground to Nevada Falls.

$1 \cdot 4 = 2 \cdot x$ The cross products are equal.
$4 = 2x$ x is multiplied by 2.
$\frac{4}{2} = \frac{2x}{2}$ Divide both sides by 2.
$2 = x$

Your child will also use ratio, comparison to 100, in solving **percent** problems that involve discounts, tips, and sales tax.

A music store has a sign that reads "10% off the regular price." If Shaundelle wants to buy a CD whose regular price is $14.99, about how much will she pay for her CD after the discount?

Step 1: First round $14.99 to $15.

Step 2: Find 10% of $15 by multiplying 0.10 • $15.

10% of 15 = 0.10 • $15 = $1.50 (Moving the decimal point one place to the left is a shortcut.)

The approximate discount is $1.50. Subtract this amount from $15.00 to find the approximate cost of the CD.

$15.00 − $1.50 = $13.50

Shaundelle will pay about $13.50 for the CD.

For additional resources, visit go.hrw.com and enter the keyword MR7 Parent.

Name _____ Date _____ Class _____

Practice A
LESSON 7-1 Ratios and Rates

Use the table to write each ratio.

1. angel fish to tiger barbs _____

2. red-tail sharks to clown loaches _____

3. catfish to angel fish _____

4. clown loaches to tiger barbs _____

5. catfish to red-tail sharks _____

Caroline's Pet Fish	
Tiger Barbs	5
Catfish	1
Angel fish	4
Red-tail sharks	1
Clown loaches	3

6. Write three equivalent ratios to compare the number of black triangles in the picture with the total number of triangles. _____

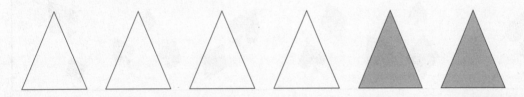

Use the table to write each ratio.

7. gray male kittens to gray female kittens

_____

8. white female kittens to white male kittens

_____

Caroline's Kittens		
	White	Gray
Male	3	2
Female	5	5

9. A candy store sells 2 ounces of chocolate for $0.80 and 3 ounces of chocolate for $0.90. How much does the store charge per ounce for the 2 ounces of chocolate? How much does the store charge per ounce for the 3 ounces of chocolate? Which is the better deal?

_____

Name _____ Date _____ Class _____

Practice B
LESSON 7-1 Ratios and Rates

Use the table to write each ratio.

1. lions to elephants _____

2. giraffes to otters _____

3. lions to seals _____

4. seals to elephants _____

5. elephants to lions _____

Animals in the Zoo	
Elephants	12
Giraffes	8
Lions	9
Seals	10
Otters	16

6. Write three equivalent ratios to compare the number of diamonds with the number of spades in the box. _____

Use the table to write each ratio as a fraction.

7. Titans wins to Titans losses _____

8. Orioles losses to Orioles wins _____

9. Titans losses to Orioles losses _____

10. Orioles wins to Titans wins _____

Baseball Team Stats		
	Titans	Orioles
Wins	12	9
Losses	14	15

11. A 6-ounce bag of raisins costs $2.46. An 8-ounce bag of raisins costs $3.20. Which is the better deal? _____

12. Barry earns $36.00 for 6 hours of yard work. Henry earns $24.00 for 3 hours of yard work. Who has the better hourly rate of pay? _____

Copyright © by Holt, Rinehart and Winston.
All rights reserved.

Holt Mathematics

Name _____ Date _____ Class _____

LESSON 7-1 Practice C
Ratios and Rates

Use the table to write each ratio.

Store T-shirt Inventory, by Color	
Red	24
Blue	42
Green	36
Purple	51
Yellow	60

1. red and blue T-shirts to green T-shirts

   _____

2. purple T-shirts to yellow and green T-shirts

   _____

3. blue and green T-shirts to purple and red T-shirts

   _____

4. red T-shirts to all other T-shirt colors

   _____

Write each ratio three different ways.

5. seven to twenty-one

6. $\frac{12}{50}$

7. 18 to 10

_____   _____   _____

_____   _____   _____

Write three equivalent ratios for each ratio.

8. 19 to 38

9. five to three

10. $\frac{20}{24}$

_____   _____   _____

11. A 12-ounce bag of birdseed costs $3.12. A 16-ounce bag of birdseed costs $3.84. Which is the better deal? How much money per ounce would you save by buying that size bag instead of the other?

    _____

    _____

12. There are 60 players on a high school football team. The ratio of juniors and seniors to freshmen and sophomores on the team is 2:3. The ratio of juniors to seniors on the team is 1:2. How many juniors are on the team? How many seniors?

    _____

Copyright © by Holt, Rinehart and Winston.
All rights reserved.

Holt Mathematics

Name _____ Date _____ Class _____

LESSON 7-1 Reteach
Ratios and Rates

A ratio is a comparison of two quantities by division.

To compare the number of times vowels are used to the number of time consonants are used in the word "mathematics," first find each quantity.

Number of times vowels are used: 4
Number of times consonants are used: 7

Then write the comparison as a ratio, using the quantities in the same order as they appear in the word expression. There are three ways to write a ratio.

$\frac{4}{7}$ 4 to 7   4:7

Write each ratio.

1. days in May to days in a year

2. sides of triangle to sides of a square

Equivalent ratios are ratios that name the same comparison.

The ratio of inches in a foot to inches in a yard is $\frac{12}{36}$. To find equivalent ratios, divide or multiply the numerator and denominator by the same number.

$\frac{12}{36} = \frac{12 \div 3}{36 \div 3} = \frac{4}{12}$ $\frac{12}{36} = \frac{12 \cdot 2}{36 \cdot 2} = \frac{24}{72}$

So, $\frac{12}{36}$, $\frac{4}{12}$, and $\frac{24}{72}$ are equivalent ratios.

Write three equivalent ratios to compare each of the following.

3. 8 triangles to 12 circles

4. 20 pencils to 25 erasers

5. 5 girls to 6 boys

6. 10 pants to 14 shirts

Holt Mathematics

LESSON 7-1

Reteach
Ratios and Rates (continued)

A rate is a comparison of two quantities that have different units of measure.

Suppose a bus travels 150 miles in 3 hours. The rate could be written as $\frac{150 \text{ miles}}{3 \text{ hours}}$.

When the second term of a rate is 1 unit, the rate is a unit rate.

To write $\frac{150 \text{ miles}}{3 \text{ hours}}$ hours as a unit rate, divide each term by 3.

$\frac{150 \text{ miles}}{3 \text{ hours}}$

$= \frac{150 \text{ miles} \div 3}{3 \text{ hours} \div 3}$

$= \frac{50 \text{ miles}}{1 \text{ hour}}$

The unit rate is $\frac{50 \text{ miles}}{\text{hour}}$.

Find each unit rate.

7. $\frac{40 \text{ books}}{2 \text{ shelves}}$

8. $\frac{36 \text{ students}}{6 \text{ groups}}$

9. $\frac{300 \text{ seconds}}{5 \text{ minutes}}$

10. $\frac{54 \text{ miles}}{2 \text{ gallons}}$

11. $\frac{4 \text{ miles}}{20 \text{ minutes}}$

12. $\frac{\$1.29}{3 \text{ pounds}}$

13. $\frac{72 \text{ hours}}{3 \text{ days}}$

14. $\frac{42 \text{ trading cards}}{6 \text{ packs}}$

Holt Mathematics

Name _____ Date _____ Class _____

LESSON 7-1 Challenge
The Golden Ratio

For centuries, people all over the world have considered a certain rectangle to be one of the most beautiful shapes. Which of these rectangles do you find the most attractive?

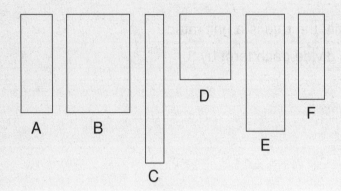

If you are like most people, you chose rectangle B. Why? It's a golden rectangle, of course! In a golden rectangle, the ratio of the length to the width is called the **golden ratio**—about 1.6 to 1.

The golden ratio pops up all over the place—in music, sculptures, the Egyptian pyramids, seashells, paintings, pinecones, and of course in rectangles.

To create your own golden rectangle, just write a ratio equivalent to the golden ratio. This will give you the length and width of another golden rectangle.

Golden Ratio

$$\frac{\ell}{w} = \frac{1.6}{1}$$

$w = 1$ in.

$\ell = 1.6$ in.

Use a ruler to draw a new golden rectangle in the space below. Then draw several non-golden rectangles around it. Now conduct a survey of your family and friends to see if they choose the golden rectangle as their favorite.

Name _____ Date _____ Class _____

LESSON 7-1 Problem Solving
Ratios and Rates

Use the table to answer each question.

Atomic Particles of Elements

Element	Protons	Neutrons	Electrons
Gold	79	118	79
Iron	26	30	26
Neon	10	10	10
Platinum	78	117	78
Silver	47	61	47
Tin	50	69	50

1. What is the ratio of gold protons to silver protons?

2. What is the ratio of gold neutrons to platinum protons?

3. What are two equivalent ratios of the ratio of neon protons to tin protons?

4. What are two equivalent ratios of the ratio of iron protons to iron neutrons?

Circle the letter of the correct answer.

5. A ratio of one element's neutrons to another element's electrons is equivalent to 3 to 5. What are those two elements?
   A iron neutrons to tin electrons
   B gold neutrons to tin electrons
   C tin neutrons to gold electrons
   D neon neutrons to iron electrons

6. The ratio of two elements' protons is equivalent to 3 to 1. What are those two elements?
   F gold to tin
   G neon to tin
   H platinum to iron
   J silver to gold

7. Which element in the table has a ratio of 1 to 1, no matter what parts you are comparing in the ratio?
   A iron   C tin
   B neon   D silver

8. If the ratio for any element is 1:1, which two parts is the ratio comparing?
   F protons to neutrons
   G electrons to neutrons
   H protons to electrons
   J neutrons to electrons

Name _____ Date _____ Class _____

LESSON 7-1 Reading Strategies
Use the Context

A **ratio** is a comparison between two similar quantities. The picture below shows geometric figures. You can write ratios to compare the figures.

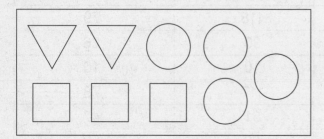

Compare the number of triangles to the total number of figures. This comparison can be written as a ratio in three different ways.

$\dfrac{\text{number of triangles}}{\text{total figures}}$ → $\dfrac{2}{9}$ Read: "two to nine."

2 to 9

2:9   Read: "two to nine."

Compare the number of squares to the number of circles.

1. Write the ratio that compares the number of squares to the number of circles in three different ways.

_____

A **rate** compares two different kinds of quantities. Rates can be shown in different ways.

You can buy 3 cans of juice for $4. The comparison of juice to money can be written:

$\dfrac{3 \text{ cans}}{\$4}$ → $\dfrac{3}{4}$ 3 to 4   3:4

Julie can jog eight miles in two hours. Use this information to complete Exercises 2–4.

2. Write the rate using words. _____

3. Write the rate with numbers in three different ways.

_____

4. Compare ratios and rates. How are they alike?

Name _____ Date _____ Class _____

Puzzles, Twisters & Teasers
LESSON 7-1 Cool Runner!

Fill in the crossword puzzle. Unscramble the circled letters to solve the riddle.

Down

1. A comparison to 1 of something is a _____ rate.

2 & 4. Hector earns $24 for 3 hours of work, while Myrna earns $17 for 2 hours. We would say Myrna's rate is a
   ____2____ ____4____.

3. A comparison of two quantities using division.

6. The ratio of boys to girls is 5:1. Number of girls if there are 10 boys.

Across

5. _____ ratios are different ratios that name the same comparison.

7. The number of boys when the ratio of boys to girls is 10:3 and the number of girls is 3.

8. The middle word when you read 10:7 ("ten _____ seven").

9. A _____ compares two quantities that have different units of measure.

What runs but never gets out of breath? ___ ___ ___ ___ ___

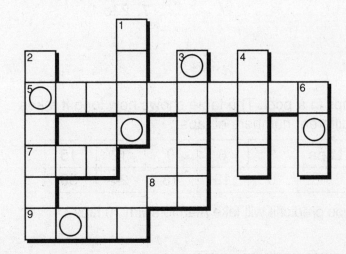

Name _____ Date _____ Class _____

LESSON 7-2 Practice A
Using Tables to Explore Equivalent Ratios and Rates

Use each table to find three equivalent ratios.

1. $\frac{1}{5}$

1			
5			

Equivalent ratios: $\frac{1}{5}$, _____, _____, and _____

2. 3 to 8

3			
8			

Equivalent ratios: 3 to 8, _____, _____, and _____

3. 80:40

80			
40			

Equivalent ratios: 80:40, _____, _____, and _____

Use a table to find three equivalent ratios.

4. 3 to 6

5. $\frac{7}{3}$

6. 1:2

7. 2 to 1

8. Alan swims laps in a pool. The table shows how long it takes him to swim different numbers of laps.

Number of Laps	3	6	9	12	15
Time (min)	6	12	18	24	30

How long do you predict it will take Alan to swim 10 laps?

Name _____ Date _____ Class _____

LESSON 7-2 Practice B
Using Tables to Explore Equivalent Ratios and Rates

Use a table to find three equivalent ratios.

1. 4 to 7

2. $\frac{10}{3}$

3. 2:5

4. 8 to 9

5. 3 to 15

6. $\frac{30}{90}$

7. 1:3

8. $\frac{7}{2}$

9. Britney does sit-ups every day. The table shows how long it takes her to do different numbers of sit-ups.

Number of Sit-Ups	10	30	50	200	220
Time (min)	2	6	10	40	44

How long do you predict it will take Britney to do 120 sit-ups?

_____

10. The School Supply Store has markers on sale. The table shows some sale prices.

Number of Markers	12	8	6	4	2
Cost ($)	9.00	6.00	4.50	3.00	1.50

How much do you predict you would pay for 10 markers?

_____

Practice C
LESSON 7-2: Using Tables to Explore Equivalent Ratios and Rates

Use a table to find three equivalent ratios.

1. 5 to 11

2. $\frac{17}{19}$

3. 8:7

4. 6 to 13

5. 36 to 12

6. $\frac{48}{90}$

Multiply and divide each ratio to find two equivalent ratios.

7. 10:20

8. $\frac{6}{9}$

9. $\frac{118}{66}$

10. 25:100

11. Spring Street Middle School orders 9 calculators for every 12 students. The table shows how many calculators the school orders for certain numbers of students.

Students	12	36	60	96	240
Calculators	9	27	45	72	180

How many calculators do you predict the school would order for 132 students?

Reteach

7-2 Using Tables to Explore Equivalent Ratios and Rates

You can use a table to find ratios equivalent to $\frac{1}{4}$.
Write the numerator in the top box for the original ratio.
Write the denominator in the bottom box for the original ratio.
Then multiply the numerator and the denominator by 2, 3, and 4.

Original ratio	1·2	1·3	1·4
1	2	3	4
4	8	12	16
	4·2	4·3	4·4

Use each new numerator and denominator to write an equivalent ratio.

So, the ratios $\frac{2}{8}$, $\frac{3}{12}$, and $\frac{4}{16}$ are equivalent to $\frac{1}{4}$.

---

1. Use the table to find three ratios equivalent to $\frac{2}{5}$.

2			
5			

Equivalent ratios: $\frac{2}{5}$, _____, _____, and _____

You can use a table to find ratios equivalent to 3 to 8.
Write the first number in the top box for the original ratio.
Write the second number in the bottom box for the original ratio.
Then multiply both numbers by 2, 3, and 4.

Original ratio	3·2	3·3	3·4
3	6	9	12
8	16	24	32
	8·2	8·3	8·4

Use each new top number and bottom number to write an equivalent ratio.

So, the ratios 6 to 16, 9 to 24, and 12 to 32 are equivalent to 3 to 8.

---

2. Use the table to find three ratios equivalent to 4 to 10.

4			
10			

Equivalent ratios: 4 to 10, _____, _____, and _____

Name _____ Date _____ Class _____

LESSON 7-2 Challenge
It's All Black and White!

This grid has a black-to-white ratio of 5 to 4.

Use the black-to-white ratio to make groups of grids.
Then complete the table of equivalent ratios.

Black	5	10	15							
White	4									

Name _____ Date _____ Class _____

LESSON 7-2 Problem Solving
Using Tables to Explore Equivalent Ratios and Rates

Use the table to answer the questions.

School Outing Student-to-Parent Ratios

Number of Students	8	16	24	32	40	48	56	64	72
Number of Parents	2	4	6	8	10	12	14	16	18

1. Each time some students go on a school outing, their teachers invite students' parents to accompany them. Predict how many parents will accompany 88 students.

2. Next week 112 students will go to the Science Museum. Their teachers invited some of the students' parents to go with them. How many parents do you predict will go with the students to the Science Museum?

Circle the letter of the correct answer.

3. Tanya's class of 28 students will be going to the Nature Center. How many parents do you predict Tanya's teacher will invite to accompany them?
   A 5 parents
   B 7 parents
   C 9 parents
   D 11 parents

4. Some students will be going on an outing to the local police station. Their teachers invited 13 parents to accompany them. How many students do you predict will be going on the outing?
   F 49 students
   G 50 students
   H 51 students
   J 52 students

5. In June, all of the students in the school will be going on their annual picnic. If there are 416 students in the school, what do you predict the number of parents accompanying them on the picnic will be?
   A 52 parents
   B 78 parents
   C 104 parents
   D 156 parents

6. On Tuesday, all of the sixth-grade students will be going to the Space Museum. Their teachers invited 21 parents to accompany them. How many sixth graders do you predict will be going to the Space Museum?
   F 80 sixth graders
   G 82 sixth graders
   H 84 sixth graders
   J 86 sixth graders

Holt Mathematics

Name _____ Date _____ Class _____

LESSON 7-2 Reading Strategies
Understand Vocabulary

Equivalent ratios are ratios that name the same comparison. The box below shows different ratios. You can find equivalent ratios by multiplying and dividing. Then you can organize them in a table.

| $\frac{3}{2}$ | 6 to 4 | $\frac{18}{10}$ | 15:10 | 12:8 |

Look for equivalent ratios. Start with $\frac{3}{2}$. Multiply the numerator and denominator by 2.

$\frac{3}{2} = \frac{3 \cdot 2}{2 \cdot 2} = \frac{6}{4}$

The resulting ratio is $\frac{6}{4}$. So 6 to 4 is equivalent to $\frac{3}{2}$.

Try $\frac{18}{10}$. Divide the numerator and denominator by 6.

$\frac{18}{10} = \frac{18 \div 6}{10 \div 6} = \frac{3}{1.7}$

The resulting ratio is not $\frac{3}{2}$. So $\frac{18}{10}$ is not equivalent to $\frac{3}{2}$.

Try 15:10. Divide each number by 5.
$15 \div 5 = 3$
$10 \div 5 = 2$
The resulting ratio is 3:2. So 15:10 is equivalent to $\frac{3}{2}$.

Try 12:8. Divide each number by 4.
$12 \div 4 = 3$
$8 \div 4 = 2$
The resulting ratio is 3:2. So 12:8 is equivalent to $\frac{3}{2}$.

Organize the equivalent ratios in a table. Write the ratios in order from least terms to greatest terms.

3	6	12	15
2	4	8	10

1. Find the equivalent ratios in the box.

| $\frac{25}{35}$ | 5 to 7 | 15:21 | 10 to 15 | $\frac{50}{70}$ |

Equivalent ratios: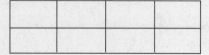

2. Organize the equivalent ratios in the table in order from least terms to greatest terms.

Puzzles, Twisters, and Teasers
LESSON 7-2 *Unlike the Others*

Each row of problems has 3 equivalent ratios and 1 that is not. Circle the one that is not equivalent to the others. Write the circled letters in the corresponding spaces to solve the riddle.

1. $\frac{5}{10}$ **S**   20:30 **N**   10 to 20 **M** $\frac{1}{2}$ **P**

2. 24:36 **U** $\frac{12}{16}$ **C**   6:9 **Z**   48 to 72 **I**

3. $\frac{96}{120}$ **F**   24 to 30 **A**   12:15 **X** $\frac{5}{4}$ **L**

4. 88 to 100 **D**   22 to 25 **B**   116:200 **E** $\frac{264}{300}$ **O**

5. 9 to 5 **C** $\frac{81}{40}$ **E**   99:55 **K**   135 to 75 **Y**

No sooner spoken than broken. What is it?

S I __ __ __ __ __
    3 5 1 2 4

LESSON 7-3 Practice A
Proportions

Find the missing value in each proportion.

1. $\frac{1}{2} = \frac{n}{6}$

2. $\frac{6}{9} = \frac{n}{3}$

3. $\frac{n}{14} = \frac{2}{7}$

_____   _____   _____

4. $\frac{2}{3} = \frac{6}{n}$

5. $\frac{n}{5} = \frac{12}{15}$

6. $\frac{2}{n} = \frac{1}{6}$

_____   _____   _____

7. $\frac{10}{2} = \frac{n}{4}$

8. $\frac{1}{4} = \frac{2}{n}$

9. $\frac{16}{8} = \frac{n}{4}$

_____   _____   _____

Write a proportion for each model.

10.

_____

11.

_____

12. Jeff made 2 out of every 5 baskets he shot during basketball practice. If he took 25 shots, how many baskets did he make?

_____

13. Tyra gets 2 quarters for every 3 newspapers she delivers. If she delivers 21 newspapers, how many quarters will she get? How much money is that in all?

_____

Name _____ Date _____ Class _____

LESSON 7-3
Practice B
Proportions

Find the missing value in each proportion.

1. $\dfrac{24}{8} = \dfrac{n}{2}$

2. $\dfrac{4}{9} = \dfrac{20}{n}$

3. $\dfrac{n}{36} = \dfrac{5}{6}$

_____   _____   _____

4. $\dfrac{n}{5} = \dfrac{4}{10}$

5. $\dfrac{3}{9} = \dfrac{2}{n}$

6. $\dfrac{6}{n} = \dfrac{3}{7}$

_____   _____   _____

7. $\dfrac{5}{3} = \dfrac{n}{6}$

8. $\dfrac{9}{6} = \dfrac{6}{n}$

9. $\dfrac{2}{130} = \dfrac{1}{n}$

_____   _____   _____

Write a proportion for each model.

10.

_____

11.

_____

12. Shane's neighbor pledged $1.25 for every 0.5 miles that Shane swims in the charity swim-a-thon. If Shane swims 3 miles, how much money will his neighbor donate?

_____

13. Barbara's goal is to practice piano 20 minutes for every 5 minutes of lessons she takes. If she takes a 20 minute piano lesson this week, how many minutes should she practice this week?

_____

Name _____ Date _____ Class _____

LESSON 7-3 Practice C
Proportions

Find the missing value in each proportion.

1. $\dfrac{6}{15} = \dfrac{n}{45}$

2. $\dfrac{n}{160} = \dfrac{1}{40}$

3. $\dfrac{2}{8} = \dfrac{n}{56}$

_____   _____   _____

4. $\dfrac{13}{26} = \dfrac{n}{4}$

5. $\dfrac{4}{9} = \dfrac{32}{n}$

6. $\dfrac{n}{16} = \dfrac{14}{32}$

_____   _____   _____

7. $\dfrac{1}{17} = \dfrac{0.5}{n}$

8. $\dfrac{8.1}{9} = \dfrac{n}{15}$

9. $\dfrac{9.1}{7} = \dfrac{n}{5}$

_____   _____   _____

10. Use circles and triangles to draw a model for the proportion $\dfrac{5}{6} = \dfrac{10}{12}$.

11. Use hearts and diamonds to draw a model for the proportion $\dfrac{3}{4} = \dfrac{9}{12}$.

12. To avoid dehydration, a person should drink 8 ounces of water for every 15 minutes of exercise. How much water should Hahn drink if he cycles for 135 minutes?

_____

13. Leo has entered a reading contest to raise money for charity. His aunt has agreed to pay Leo $0.13 for every 5 pages that he reads. Leo's uncle has promised to match every whole dollar that Leo collects in the contest with $1.75. If Leo reads 365 pages, how much money will his aunt donate to the charity? How much will Leo's uncle give to match the aunt's donation?

Reteach
7-3 Proportions

A proportion is an equation that shows two equivalent ratios.
$\frac{3}{4} = \frac{9}{12}$ is an example of a proportion.
$3 \cdot 12 = 36$ and $4 \cdot 9 = 36$.   The cross products of proportions are equal.

You can use cross products to find the missing value in a proportion.

$$\frac{3}{x} = \frac{12}{48}$$

$12 \cdot x = 3 \cdot 48$ To find x, first find the cross products.

$12x = 144$

Think: $144 \div 12 = x$ Then use a related math sentence to
$x = 12$ solve the equation.

So, $\frac{3}{12} = \frac{12}{48}$.

Find the cross products to solve each proportion.

1. $\frac{x}{8} = \frac{3}{4}$

   $x \cdot 4 =$ _____

   _____

2. $\frac{2}{3} = \frac{x}{6}$

   $2 \cdot 6 =$ _____

   _____

3. $\frac{2}{5} = \frac{4}{x}$

   $2 \cdot x =$ _____

   _____

4. $\frac{6}{x} = \frac{1}{3}$

   $6 \cdot 3 =$ _____

   _____

5. $\frac{3}{8} = \frac{12}{x}$

6. $\frac{3}{5} = \frac{6}{x}$

7. $\frac{x}{8} = \frac{2}{16}$

8. $\frac{2}{9} = \frac{4}{x}$

9. $\frac{3}{4} = \frac{15}{x}$

10. $\frac{1}{2} = \frac{x}{30}$

11. $\frac{x}{5} = \frac{24}{30}$

12. $\frac{25}{35} = \frac{5}{x}$

Challenge

LESSON 7-3 Patriotic Proportions

On August 21, 1959, President Eisenhower signed an order that established the official proportions of the United States flag. No matter what size the flag is, it must match those proportions to be used officially.

Official Proportions for the United States Flag	
Width of flag	1
Length of flag	$1\frac{9}{10}$
Width of union	$\frac{7}{13}$
Length of union	$\frac{19}{25}$
Width of each stripe	$\frac{1}{13}$

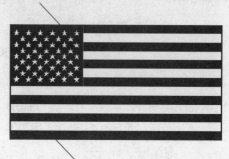

The union is the blue area.
The 50 stars represent the 50 states.

The 13 stripes represent the first 13 states.

Use the official proportions to find the missing dimension of each flag.

1. Length of flag = 10 feet; Width of flag = _____

2. Width of flag = 57 yards; Length of flag = _____

3. Width of flag = 13 centimeters; Width of Union = _____

4. Width of flag = 260 inches; Width of each stripe = _____

5. Length of flag = 25 meters; Length of Union = _____

Choose a width in inches for a United States flag. Then use a ruler to draw your flag with the official proportional length in the space below.

Problem Solving
7-3 Proportions

Write the correct answer.

1. For most people, the ratio of the length of their head to their total height is 1:7. Use proportions to test your measurements and see if they match this ratio.

2. The ratio of an object's weight on Earth to its weight on the Moon is 6:1. The first person to walk on the Moon was Neil Armstrong. He weighed 165 pounds on Earth. How much did he weigh on the Moon?

3. It has been found that the distance from a person's eye to the end of the fingers of his outstretched hand is proportional to the distance between his eyes at a 10:1 ratio. If the distance between your eyes is 2.3 inches, what should the distance from your eye to your outstretched fingers be?

4. Chemists write the formula of ordinary sugar as $C_{12}H_{22}O_{11}$, which means that the ratios of 1 molecule of sugar are always 12 carbon atoms to 22 hydrogen atoms to 11 oxygen atoms. If there are 4 sugar molecules, how many atoms of each element will there be?

Circle the letter of the correct answer.

5. A healthy diet follows the ratio for meat to vegetables of 2.5 servings to 4 servings. If you eat 7 servings of meat a week, how many servings of vegetables should you eat?
   A 28 servings   C 14 servings
   B 17.5 servings D 11.2 servings

6. A 150-pound person will burn 100 calories while sitting still for 1 hour. Following this ratio, how many calories will a 100-pound person burn while sitting still for 1 hour?
   F $666\frac{2}{3}$ calories   H $6\frac{2}{3}$ calories
   G $66\frac{2}{3}$ calories   J 6 calories

7. Recently, 1 U.S. dollar was worth 1.58 in euros. If you exchanged $25 at that rate, how many euros would you get?
   A 39.50 euros
   B 15.82 euros
   C 26.58 euros
   D 23.42 euros

8. Recently, 1 U.S. dollar was worth 0.69 English pound. If you exchanged 500 English pounds, how many dollars would you get?
   F 345 U.S. dollars
   G 725 U.S. dollars
   H 500.69 U.S dollars
   J 499.31 U.S. dollars

Name _____ Date _____ Class _____

LESSON 7-3 Reading Strategies
Use Graphic Aids

A **proportion** is a statement of two equal ratios. This statement is written as an equation.

One cup of juice contains 50 calories.

This statement can be written as a ratio.
$\dfrac{\text{cups}}{\text{calories}} \rightarrow \dfrac{1}{50}$

Two cups of juice contain 100 calories.

This statement can also be written as a ratio. $\dfrac{\text{cups}}{\text{calories}} \rightarrow \dfrac{2}{100}$

Are these two ratios equal?

Step 1: Write a proportion with the two ratios.

$\dfrac{1}{50} = \dfrac{2}{100}$ → Read: "1 is to 50 as 2 is to 100."

Step 2: Find the cross products. If cross products are equal, the ratios are equal and form a proportion.

$\dfrac{1}{50} \times \dfrac{2}{100}$ $2 \times 50 = 100$
$1 \times 100 = 100$

Use this picture to answer the questions.

1. What is the ratio of striped circles to total circles? _____

2. What is the ratio of black circles to white circles? _____

3. Find the cross products. Write = or ≠ to complete.

4. Do $\dfrac{2}{8}$ and $\dfrac{1}{4}$ form a proportion?

Name _____ Date _____ Class _____

Puzzles, Twisters & Teasers
LESSON 7-3 *Too Much!*

Solve the problems and circle your answers.

Using the letters next to your answers, create three words that mean different things, but are all pronounced the same! You will use two of the letters more than once.

1. Chung is giving medicine to his cat Princess. The bottle recommends 3 pills for a 15 pound cat, but Princess weighs only 10 pounds. How many pills should Chung give?

    R 1 pill   **W** 2 pills   **A** 3 pills

2. Find the missing value: $\frac{5}{n} = \frac{15}{21}$

    O 7   **D** 3   **M** 5

3. Suri knows that she needs to study about twenty minutes a night for each hour class in math, and about thirty minutes for each hour class in history. Normally she has one hour of each class every day. But today she had math class for an hour and a half and only a half-hour history class. Will her homework take more, less, or the same amount of time tonight?

    F more   **P** same   **T** less

Now use the letters next to your answers to figure out the three words.

A number: ___ ___ ___

Also: ___ ___ ___

A preposition: ___ ___

Name _____ Date _____ Class _____

LESSON 7-4 Practice A
Similar Figures

Tell whether the figures in each pair are similar.

1.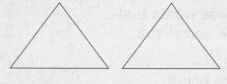

   _____

2.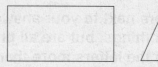

   _____

3.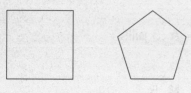

   _____

4.

   _____

5. The two triangles are similar. Find the missing length x and the measure of $\angle F$.

   _____

   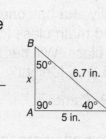 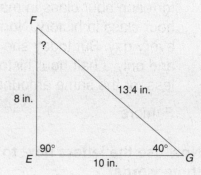

6. The two triangles are similar. Find the missing length m and the measure of $\angle O$.

   _____

   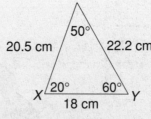 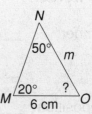

7. Two rectangular photos are similar. The larger photo is 6 inches wide and 8 inches long. The smaller photo is 3 inches wide. What is the smaller photo's length?

   _____

8. Two triangular mirrors are similar. The first mirror's angles all measure 60°. What are the measures of the second mirror's angles? Explain how you know.

   _____

Name _____ Date _____ Class _____

LESSON 7-4 Practice B
Similar Figures

Write the correct answers.

1. The two triangles are similar. Find the missing length x and the measure of ∠A. _____

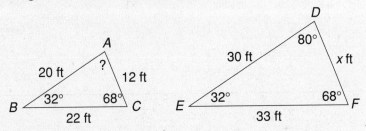

2. The two triangles are similar. Find the missing length x and the measure of ∠J. _____

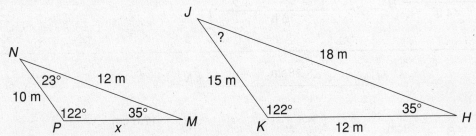

3. The two triangles are similar. Find the missing length x and the measure of ∠N. _____

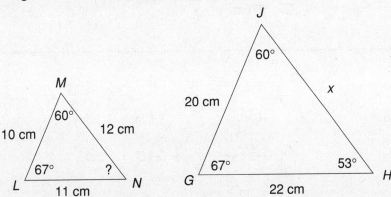

4. Juanita planted two flower gardens in similar square shapes. What are the measures of all the angles in each garden? Explain how you know.

_____

_____

Copyright © by Holt, Rinehart and Winston.
All rights reserved.

Holt Mathematics

Name _____ Date _____ Class _____

LESSON 7-4 Practice C
Similar Figures

The figures in each pair are similar. Find the unknown measures.

1.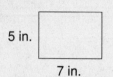

2.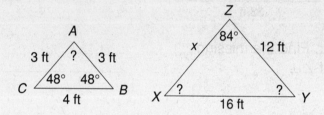

3.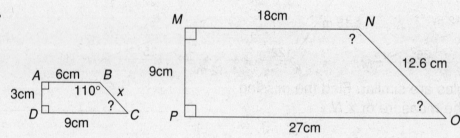

4.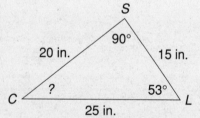

5. Two regular pentagons are similar. One side of the first pentagon is 3 m long, and the perimeter of the second pentagon is three times as long as the first pentagon. What are the lengths of each side of the second pentagon? _____

6. A 7-by-9 foot rectangle is similar to a second rectangle whose perimeter is 260 ft. What are the dimensions of the second rectangle? _____

LESSON 7-4 Reteach
Similar Figures

Two figures are similar if they have the same shape but are different sizes.

Similar figures have corresponding sides and corresponding angles. Corresponding sides are proportional. Corresponding angles are congruent.

Look at the similar triangles below.

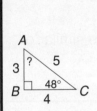 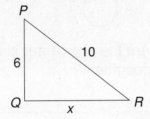

$\overline{AB}$ corresponds to $\overline{PQ}$.   ∠A corresponds to ∠P.
$\overline{BC}$ corresponds to $\overline{QR}$.   ∠B corresponds to ∠Q.
$\overline{AC}$ corresponds to $\overline{PR}$.   ∠C corresponds to ∠R.

What is the length of $\overline{QR}$?

$\dfrac{AB}{BC} = \dfrac{PQ}{QR}$ Set up a proportion.

$\dfrac{3}{4} = \dfrac{6}{x}$ Substitute the values.

$3 \cdot x = 4 \cdot 6$ The cross products are equal.

$3x = 24$ x is multiplied by 3.

$\dfrac{3x}{3} = \dfrac{24}{3}$ Divide both sides by 3.

$x = 8$

So, the length of $\overline{QR}$ is 8 units.

Find each missing length.

1.

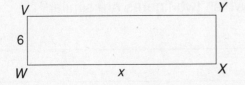

2.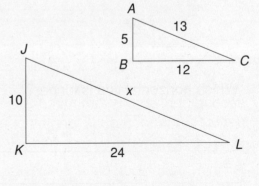

Name _____ Date _____ Class _____

LESSON 7-4 Challenge
You Won't Believe Your Eyes!

Answer each question by looking at the drawings below. Then use what you know about similar and congruent figures to verify your answers.

1.

   Are the two line segments congruent?

   _____

2.

   Are the two center circles similar or congruent?

   _____

3.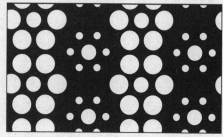

   Are any of these circles similar?

   _____

4.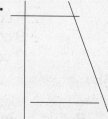

   Are any of these line segments congruent?

   _____

5.

   Which horizontal line is longer?

   _____

6.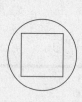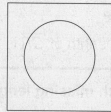

   Which two figures are congruent?
   Which two figures are similar?

   _____

Name _____ Date _____ Class _____

LESSON 7-4 Problem Solving
Similar Figures

Write the correct answer.

1. The map at right shows the dimensions of the Bermuda Triangle, a region of the Atlantic Ocean where many ships and airplanes have disappeared. If a theme park makes a swimming pool in a similar figure, and the longest side of the pool is 0.5 mile long, about how long would the other sides of the pool have to be?

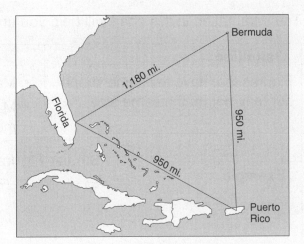

2. Completed in 1883, *The Battle of Gettysburg* is 410 feet long and 70 feet tall. A museum shop sells a print of the painting that is similar to the original. The print is 2.05 feet long. How tall is the print?

3. *Panorama of the Mississippi* was 12 feet tall and 5,000 feet long! If you wanted to make a copy similar to the original that was 2 feet tall, how many feet long would the copy have to be?

Circle the letter of the correct answer.

4. Two tables shaped like triangles are similar. The measure of one of the larger table's angles is 38°, and another angle is half that size. What are the measures of all the angles in the smaller table?

   A 19°, 9.5°, and 61.5°
   B 38°, 19°, and 123°
   C 38°, 38°, and 104°
   D 76°, 38°, and 246°

5. Two rectangular gardens are similar. The area of the larger garden is 8.28 m², and its length is 6.9 m. The smaller garden is 0.6 m wide. What is the smaller garden's length and area?

   F length = 6.9 m; area = 2.07 m²
   G length = 3.45 m; area = 4.14 m²
   H length = 3.45 m; area = 1.97 m²
   J length = 3.45 m; area = 2.07 m²

6. Which of the following is not always true if two figures are similar?

   A They have the same shape.
   B They have the same size.
   C Their corresponding sides have proportional lengths.
   D Corresponding angles are congruent.

7. Which of the following figures are always similar?

   F two rectangles
   G two triangles
   H two squares
   J two pentagons

Name _____ Date _____ Class _____

LESSON 7-4 Reading Strategies
Graphic Organizer

The information in this chart will help you understand **similar figures**.

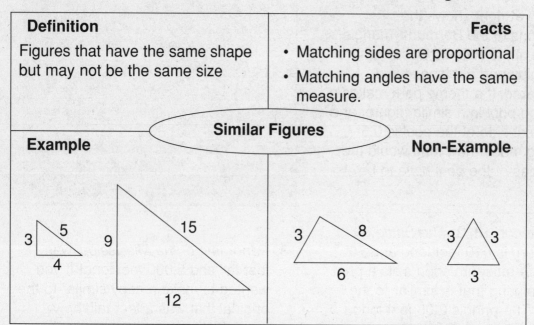

Use the chart to help you answer these questions. Write *yes*, *no*, or *maybe*.

1. Do similar figures have the same shape?
   _____

2. Are similar figures the same size?
   _____

3. Would a small square and a large square have different angle measurements?
   _____

4. Would a small square and a large square have matching sides that are proportional?
   _____

5. Would a large square and a small square be examples of similar figures?
   _____

6. Would a large square and a large triangle be similar figures?
   _____

Name _____ Date _____ Class _____

Puzzles, Twisters & Teasers
LESSON 7-4 With One Blow!

Solve each of the problems below and circle your answer. Transfer the matching letters, in order, on to the blanks to solve the riddle.

1. A small triangle has a hypotenuse of 5, and sides of 3 and 4. A larger, similar, triangle has a hypotenuse of 30. Find the lengths of the other two sides of the larger triangle.

   K 25   G 24   F 12
   W 15   S 20   I 18

2. A large parallelogram has angles of 120 degrees and 60 degrees. What are the corresponding angles of a smaller, similar parallelogram?

   A 120   V 90   R 180
   P 30   E 360   N 60

3. Alok went to the photography store to develop some film. He has three choices of sizes for his prints. He thinks that two of the sizes make similar rectangles. Which size does *not* make a rectangle similar to the other two?

   M 4 by 6   T 8 by 10   B 8 by 12

What kind of ant can break a picnic table with one blow?

A ___ ___ ___ ___ ___

Name _____ Date _____ Class _____

Practice A
LESSON 7-5 Indirect Measurement

Write the correct answer.

1. Use similar triangles to find the height of the lamppost. _____

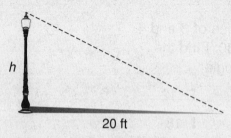

2. Use similar triangles to find the height of the man. _____

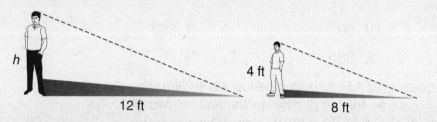

3. A 3-foot-tall boy looks into a mirror at the county fair. The mirror makes a person appear shorter. The boy appears to be 1 foot tall in the mirror. If a man appears to be 2 feet tall in the mirror, what is his actual height?

_____

4. On a sunny day, a carnation casts a shadow that is 20 inches long. At the same time, a 3-inch-tall tulip casts a shadow that is 12 inches long. How tall is the carnation?

_____

5. A bicycle casts a shadow that is 8 feet long. At the same time, a girl who is 5 feet tall casts a shadow that is 10 feet long. How tall is the bicycle?

_____

6. A sand castle casts a shadow that is 5 inches long. A 15-inch-tall bucket sitting next to the sand castle casts a shadow that is 3 inches long. How tall is the sand castle?

_____

7. Through a magnifying glass, a 2-centimeter-long bug looks like it is 12 centimeters long. How long would a 3-centimeter bug look in that same magnifying glass?

_____

8. In the late afternoon, a wagon casts a shadow that is 15 feet long. A boy pulling the wagon who is 4 feet tall casts a shadow that is 20 feet long. How tall is the wagon?

_____

Holt Mathematics

Name _____ Date _____ Class _____

LESSON 7-5 Practice B
Indirect Measurement

Write the correct answer.

1. Use similar triangles to find the height of the building. _____

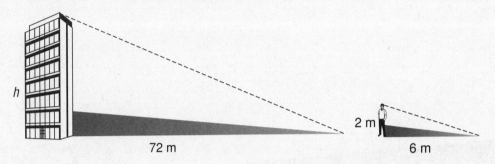

2. Use similar triangles to find the height of the taller tree. _____

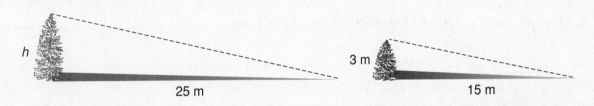

3. A lamppost casts a shadow that is 35 yards long. A 3-foot-tall mailbox casts a shadow that is 5 yards long. How tall is the lamppost?

4. A 6-foot-tall scarecrow in a farmer's field casts a shadow that is 21 feet long. A dog standing next to the scarecrow is 2 feet tall. How long is the dog's shadow?

5. A building casts a shadow that is 348 meters long. At the same time, a person who is 2 meters tall casts a shadow that is 6 meters long. How tall is the building?

6. On a sunny day, a tree casts a shadow that is 146 feet long. At the same time, a person who is 5.6 feet tall standing beside the tree casts a shadow that is 11.2 feet long. How tall is the tree?

7. In the early afternoon, a tree casts a shadow that is 2 feet long. A 4.2-foot-tall boy standing next to the tree casts a shadow that is 0.7 feet long. How tall is the tree?

8. Steve's pet parakeet is 100 mm tall. It casts a shadow that is 250 mm long. A cockatiel sitting next to the parakeet casts a shadow that is 450 mm long. How tall is the cockatiel?

Name _____ Date _____ Class _____

LESSON 7-5 Practice C
Indirect Measurement

Write the correct answer.

1. Use similar triangles to find the height of the tower. _____

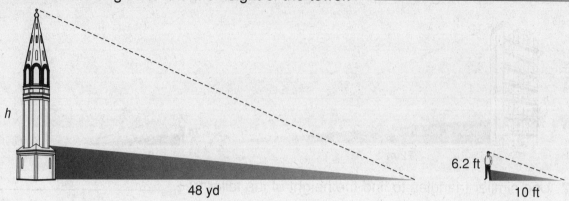

2. Use similar triangles to find the height of the man. _____

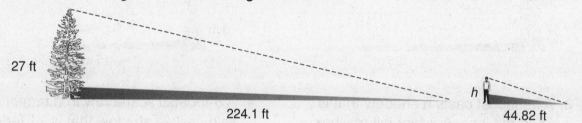

3. On a sunny day, a 6.5-foot-tall ladder casts a shadow that is 19.5 feet long. A man who is 6.2 feet tall is painting next to the ladder. How long is his shadow?

4. A building casts a shadow that is 1,125 meters long. A woman standing next to the building casts a shadow that is 6.25 meters long. She is 2.5 meters tall. How tall is the building?

5. Brian, who is twice as tall as Cole, is 6.5 feet tall. Cole casts a shadow that is 22.75 feet long. If Brian is standing next to Cole, how long is Brian's shadow?

6. A 4.5-foot-tall boy stands so the top of his shadow is even with the top of a flagpole's shadow. If the flagpole's shadow is 34 feet long, and the boy is standing 25 feet away from the flagpole, how tall is the flagpole?

7. A mother giraffe is 18.7 feet tall. Her baby is 5.25 feet tall. The baby giraffe casts a shadow that is 35.7 feet long. How long is the mother giraffe's shadow?

8. A shorter flagpole casts a shadow 15.3 feet shorter than the shadow of a longer pole. The taller pole is 26.5 feet tall and casts a shadow 47.7 feet long. How tall is the shorter pole?

LESSON 7-5 Reteach
Indirect Measurement

If you cannot measure a length directly, you can use indirect measurement. Indirect measurement uses similar figures and proportions to find lengths.

The small tree is 8 feet high and it casts a 12-foot shadow. The large tree casts a 36-foot shadow.

The triangles formed by the trees and the shadows are similar. So, their heights are proportional.

To find the height of the large tree, first set up a proportion. Use a variable to stand for the height of the large tree.

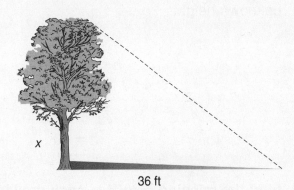

$\frac{8}{12} = \frac{x}{36}$ Write a proportion using corresponding sides.

$8 \cdot 36 = 12 \cdot x$ The cross products are equal.

$12x = 288$ x is multiplied by 12.

$\frac{12x}{12} = \frac{288}{12}$ Divide both sides by 12.

$x = 24$

So, the height of the tall tree is 24 feet.

Use indirect measurement to find the missing heights.

1.
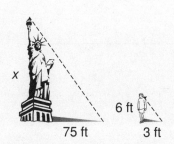

2.
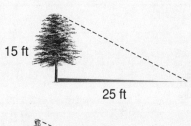

Name _____ Date _____ Class _____

LESSON 7-5 Challenge
Mirror Measurements

When it is noon, nighttime, a cloudy day, or when you are inside, there are hardly any shadows to use for indirect measurement. Instead, you can use mirrors to measure in the following way.

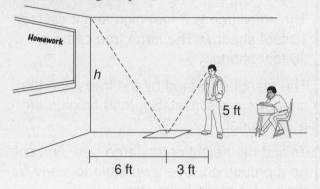

Place a mirror on the floor. Move back until you see the reflection of the top of the object you want to measure in the mirror. This creates two similar triangles. You can then use proportions to find the unknown height:

$$\frac{h}{5} = \frac{6}{3}$$
$$h \cdot 3 = 5 \cdot 6$$
$$3h = 30$$
$$\frac{3h}{3} = \frac{30}{3}$$
$$h = 10$$

So, the height of the classroom is 10 feet.

Find the missing height in each drawing to the nearest whole foot.

1.
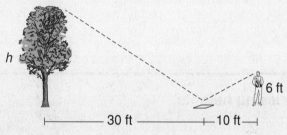

_____

2.
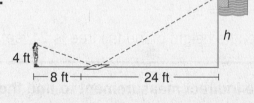

_____

3.
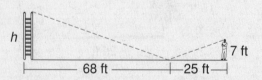

_____

4.
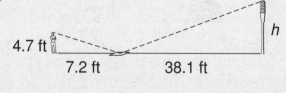

_____

Copyright © by Holt, Rinehart and Winston.
All rights reserved.

Holt Mathematics

Name _____ Date _____ Class _____

LESSON 7-5 Problem Solving
Indirect Measurement

Write the correct answer.

1. The Petronas Towers in Malaysia are the tallest buildings in the world. On a sunny day, the Petronas Towers cast shadows that are 4,428 feet long. A 6-foot-tall person standing by one building casts an 18-foot-long shadow. How tall are the Petronas Towers?

2. The Sears Tower in Chicago is the tallest building in the United States. On a sunny day, the Sears Tower casts a shadow that is 2,908 feet long. A 5-foot-tall person standing by the building casts a 10-foot-long shadow. How tall is the Sears Tower?

3. The world's tallest man cast a shadow that was 535 inches long. At the same time, a woman who was 5 feet 4 inches tall cast a shadow that was 320 inches long. How tall was the world's tallest man in feet and inches?

4. Hoover Dam on the Colorado River casts a shadow that is 2,904 feet long. At the same time, an 18-foot-tall flagpole next to the dam casts a shadow that is 72 feet long. How tall is Hoover Dam?

Circle the letter of the correct answer.

5. An NFL goalpost casts a shadow that is 170 feet long. At the same time, a yardstick casts a shadow that is 51 feet long. How tall is an NFL goalpost?
   A 100 feet
   B 56 2/3 feet
   C 10 feet
   D 1 foot

6. A gorilla casts a shadow that is 600 centimeters long. A 92-centimeter-tall chimpanzee casts a shadow that is 276 centimeters long. What is the height of the gorilla in meters?
   F 0.2 meter
   G 2 meters
   H 20 meters
   J 200 meters

7. A 6-foot-tall man casts a shadow that is 30 feet long. If a boy standing next to the man casts a shadow that is 12 feet long, how tall is the boy?
   A 2.2 feet C 2.4 feet
   B 5 feet   D 2 feet

8. An ostrich is 108 inches tall. If its shadow is 162 inches, and an emu standing next to it casts a 90-inch shadow, how tall is the emu?
   F 162 inches H 60 inches
   G 90 inches   J 194.4 inches

Name _____ Date _____ Class _____

LESSON 7-5
Reading Strategies
Following Procedures

Indirect means "not direct." We use **indirect measurement** when it is not possible to use standard measurement tools. Measuring the height of a very tall tree or water tower is difficult to do directly. An indirect method of measuring uses similar figures and proportions to find the length or height.

These two triangles are similar.

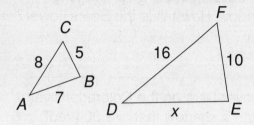

1. What side of the second triangle corresponds to side *AB*? _____

2. What side of the second triangle corresponds to side *AC*? _____

3. What side of the second triangle corresponds to side *BC*? _____

You can set up a proportion using two corresponding sides from each triangle to find the missing side.

4. Write a ratio using the lengths for sides *AC* and *DF*. _____

5. Write a ratio using the lengths for sides *AB* and *DE*. _____

6. Write the proportion for the above two ratios. _____

7. Write a ratio using the lengths for sides *BC* and *EF*. _____

8. Write a ratio using the lengths for sides *AB* and *DE*. _____

9. Write a proportion for the above two ratios. _____

Copyright © by Holt, Rinehart and Winston.
All rights reserved.

Holt Mathematics

Name _____ Date _____ Class _____

LESSON 7-5
Puzzles, Twisters & Teasers
Not Quite the Same

Solve the crossword. Unscramble the circled letters to answer the question.

Down

1. In similar _____ corresponding angles are congruent.

2. Figures that have the same shape but not necessarily the same size.

4. Solve: $\frac{5}{30} = \frac{x}{42}$

6. To sum two numbers.

7. Six, fourteen, fifty-two, and seventy-eight are all _____ numbers.

Across

3. In similar figures, corresponding _____ have lengths that are proportional.

5. Indirect _____

8. Solve: $\frac{6}{x} = \frac{18}{33}$

9. A figure with three sides is a _____-angle.

What do you use to make indirect measurements?

___ ___ ___ ___ ___ ___ ___ FIGURES

Name _____ Date _____ Class _____

LESSON 7-6

Practice A
Scale Drawings and Maps

Use the map to answer the questions.

1. On the map, the distance from Newton to Cambridge is 2 cm. What is the actual distance?

   _____

2. On the map, the distance from Arlington to Medford is 1 cm. What is the actual distance?

   _____

3. If the distance between two cities on this map measures 6 centimeters, what is the actual distance?

   _____

4. If the actual distance between two cities is 12 kilometers, how many centimeters will separate those two cities on this map?

   _____

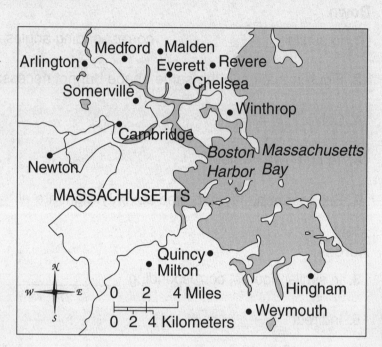

Use the scale drawing to answer each question.

5. This scale drawing is of the *Mayflower*, the ship that the first English settlers of Massachusetts used. How long was the actual *Mayflower*?

   _____

6. The height of the actual *Mayflower* was 200 feet from the bottom of the boat to the top of the tallest mast. Is the ship's height in the drawing correct?

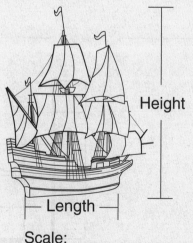

Scale:
1 inch = 100 feet

Name _____ Date _____ Class _____

LESSON 7-6 Practice B
Scale Drawings and Maps

Use the map to answer the questions.

1. On the map, the distance between Big Cypress Swamp and Lake Okeechobee is $\frac{1}{4}$ inch. What is the actual distance?

   _____

2. On the map, the distance between Key West and Cuba is $\frac{9}{10}$ inch. What is the actual distance?

   _____

3. Use a ruler to measure the distance between Key West and Key Largo on the map. What is the actual distance?

   _____

4. The Overseas Highway connects Key West to mainland Florida. It is 110 miles long. If it were shown on this map, how many inches long would it be?

   _____

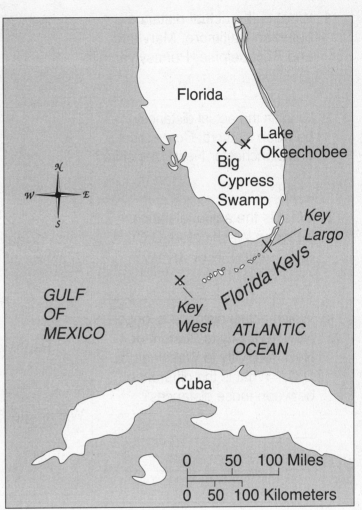

Use the scale drawing to answer each question.

5. This scale drawing is of the lighthouse on Key West, originally built in 1825. What is the actual height of the lighthouse?

   _____

6. The original lighthouse was 66 feet tall. It was rebuilt at its present height after a hurricane destroyed it in 1846. How tall would the original lighthouse be in this scale drawing?

1 inch = 40 feet

Name _____ Date _____ Class _____

LESSON 7-6 Practice C
Scale Drawings and Maps

Use a metric ruler and the map to answer the questions.

1. What is the actual distance between Baltimore, Maryland, and Philadelphia, Pennsylvania?

   _____

2. What is the actual distance between Hartford, Connecticut, and Manchester, New Hampshire?

   _____

3. What is the actual distance between Washington, D.C., and Trenton, New Jersey?

   _____

4. Which actual distance is longer: New York City to Boston, or New York City to Washington, D.C.? What is the difference between those distances?

   _____
   _____

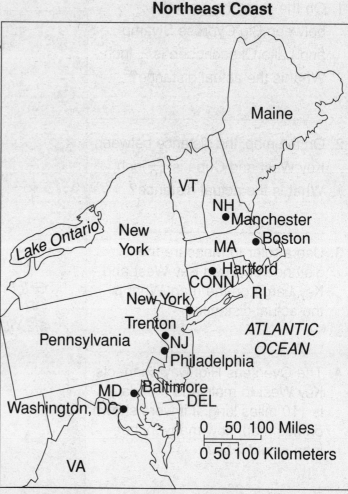

Northeast Coast

Use the scale drawing to answer each question.

5. What is the actual height of the Statue of Liberty, including its pedestal?

   _____

6. What is the actual height of the statue from the base to the tip of her torch?

   _____

7. The statue's arm that holds the torch is 42 feet long. How many inches long should it be on this drawing?

   _____

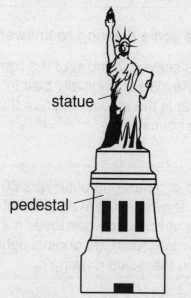

Scale: 1 inch = 100 feet

LESSON 7-6 Reteach
Scale Drawings and Maps

A scale drawing is a drawing of a real object that is proportionally smaller or larger than the real object.

A scale is a ratio between two sets of measurements. In the map below, the scale is 2 cm: 0.5 km. This means that each centimeter on the map represents 0.25 kilometer.

Store

Park

School

Library

To find the actual distance from school to the library, first measure the distance on the map using a ruler.

The distance on the map is 3 centimeters.

$\frac{2 \text{ cm}}{0.5 \text{ km}} = \frac{3 \text{ cm}}{x \text{ km}}$ Write a proportion using the scale.

$2 \cdot x = 0.5 \cdot 3$ The cross products are equal.

$2x = 1.5$ x is multiplied by 2.

$\frac{2x}{2} = \frac{1.5}{2}$ Divide both sides by 2.

$x = 0.75$

The distance from school to the library is 0.75 kilometer.

Use the map to find each actual distance.

1. from the store to the library

2. from the park to the store

3. from the park to the library

4. from the park to the school

Name _____ Date _____ Class _____

LESSON 7-6 Challenge
Solar System String

Distances in outer space are usually measured in millions of miles. Understanding or comparing such huge measurements can be difficult, and it is impossible to map or draw them in their actual scale. Here's an activity that can help you understand the vast scale of our solar system. Identify the 1-millimeter mark on your ruler. This tiny distance represents 1,000,000 miles in space! You will use it as the scale for your model: 1 millimeter = 1 million miles.

Make a scale model of our solar system.

1. Cut a piece of string 4 meters long. Tape a small piece of paper at one end of the string and label it "Sun."
2. From the sun, measure 3.6 cm. Tape a "Mercury" label there.
3. From Mercury, measure another 3.1 cm. Tape a "Venus" label there.
4. From Venus, measure another 2.6 cm. Tape an "Earth" label there.
5. From Earth, measure another 4.9 mm. Tape a "Mars" label there.
6. From Mars, measure another 34.2 cm. Tape a "Jupiter" label there.
7. From Jupiter, measure another 40.2 cm. Tape a "Saturn" label there.
8. From Saturn, measure another 89.8 cm. Tape a "Uranus" label there.
9. From Uranus, measure another 1,010 mm. Tape a "Neptune" label there.
10. From Neptune, measure another 88.1 cm. Tape a "Pluto" label there.

Now use the scale and your model to find the actual distance from Earth. For example, the distance from Earth to the sun on the string measures 93 mm, so the actual distance is 93 million miles.

Earth to Mercury: _____

Earth to Venus: _____

Earth to Mars: _____

Earth to Jupiter: _____

Earth to Saturn: _____

Earth to Uranus: _____

Earth to Neptune: _____

Earth to Pluto: _____

Name _____ Date _____ Class _____

LESSON 7-6 Problem Solving
Scale Drawings and Maps

Write the correct answer.

1. About how many kilometers long is the northern border of California along Oregon?

   _____

2. What is the distance in kilometers from Los Angeles to San Francisco?

   _____

3. How many kilometers would you have to drive to get from San Diego to Sacramento?

   _____

4. At its longest point, about how many kilometers long is Death Valley National Park?

   _____

5. Approximately what is the distance, in kilometers, between Redwood National Park and Yosemite National Park?

   _____

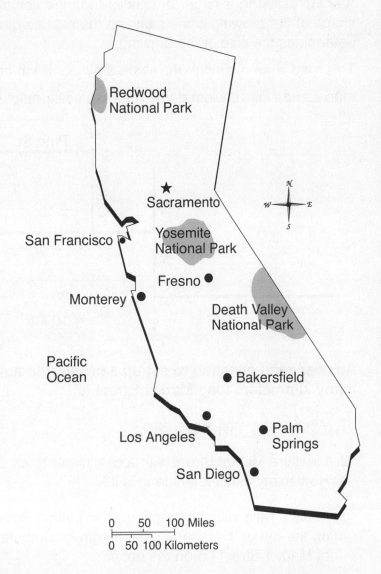

Circle the letter of the correct answer.

6. Which of the following two cities in California are about 200 kilometers apart?

   A San Diego and Los Angeles
   B Monterey and Los Angeles
   C San Francisco and Fresno
   D Palm Springs and Bakersfield

7. Joshua Tree National Park is about 200 kilometers from Sequoia National Park. How many centimeters should separate those parks on this map?

   F 110 cm   H 1 cm
   G 11 cm   J 0.11 cm

Name _____ Date _____ Class _____

LESSON 7-6 Reading Strategies
Use Graphic Aids

A **scale drawing** is larger or smaller than the actual object. The shape of the drawing is the same as the actual object. The scale determines the size of the drawing.

This map is an example of a scale drawing. Each centimeter on the map stands for 10 kilometers. The map scale ratio is $\frac{1 \text{ cm}}{10 \text{ km}}$.

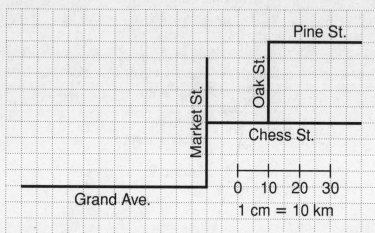

Answer each question to set up a proportion and find out how many kilometers long Market Street is.

1. What is the map scale ratio? _____

2. Measure Market Street with a centimeter ruler. How many centimeters long is it? _____

3. Make a ratio with the map length of Market Street on the top of the ratio and the length in kilometers of Market Street (x) on the bottom. _____

4. Use the ratios from Exercises 1 and 3 to write a proportion. _____

Answer each question to find out how many kilometers long Grand Avenue is.

5. How many centimeters long is Grand Avenue on the map? _____

6. Write a proportion using the map scale ratio and x divided by the map measurement of Grand Avenue. _____

Puzzles, Twisters & Teasers

7-6 In Order, Please!

On each grid below, find the actual length of the lines. (The grids have different scales.) Place the *actual* lengths in order from smallest to largest. Use the corresponding letters to solve the riddle.

Grid #1-scale: 1 unit = 5 feet

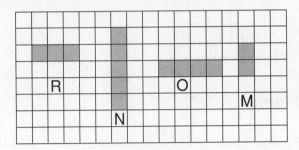

Grid #2-scale: 1 unit = 7 feet

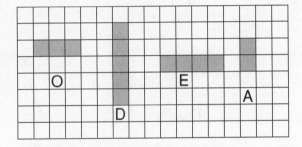

What happened when a ship carrying a load of blue paint collided with a ship carrying a load of red paint?

The crews were ___ ___ ___ ___ ___ ___ ___ ___ .

Name _____ Date _____ Class _____

LESSON 7-7 Practice A
Percents

Use the 10-by-10-square grids to model each percent.

1. 12%

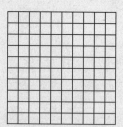

2. 67%

Write each percent as a fraction in simplest form.

3. 50%   4. 1%   5. 11%

_____   _____   _____

6. 10%   7. 99%   8. 17%

_____   _____   _____

Write each percent as a decimal.

9. 5%   10. 75%   11. 2%

_____   _____   _____

12. 15%   13. 13%   14. 90%

_____   _____   _____

15. The math workbook has 100 pages. Each chapter of the book is 10 pages long. What percent of the book does each chapter make up?

_____

16. There were 100 questions on the math test. Chi-Tang answered 88 of those questions correctly. What percent did he get correct on the test?

_____

Name _____ Date _____ Class _____

LESSON 7-7 Practice B
Percents

Write each percent as a fraction in simplest form.

1. 30%

2. 42%

3. 18%

4. 35%

5. 100%

6. 29%

7. 56%

8. 70%

9. 25%

Write each percent as a decimal.

10. 19%

11. 45%

12. 3%

13. 80%

14. 24%

15. 6%

Order the percents from least to greatest.

16. 89%, 42%, 91%, 27%

17. 2%, 55%, 63%, 31%

18. Sarah correctly answered 84% of the questions on her math test. What fraction of the test questions did she answer correctly? Write your answer in simplest form.

19. Chloe swam 40 laps in the pool, but this was only 50% of her total swimming workout. How many more laps does she still need to swim?

Name _____ Date _____ Class _____

LESSON 7-7
Practice C
Percents

Write each percent as a fraction or mixed number in simplest form.

1. 68%

2. 98%

3. 55%

_____ _____ _____

4. 84%

5. 16%

6. 60%

_____ _____ _____

7. 125%

8. 150%

9. 140%

_____ _____ _____

Write each percent as a decimal.

10. 0.5%

11. 0.25%

12. 127%

_____ _____ _____

13. 205%

14. 1165%

15. 0.08%

_____ _____ _____

Order from least to greatest.

16. 92%, 0.86, 47%, and $\frac{14}{25}$

17. 5%, $\frac{7}{100}$, 0.8%, 0.003

_____ _____

18. Of all the students who voted for their favorite ice cream, $\frac{1}{2}$ chose chocolate and $\frac{2}{5}$ chose vanilla. What percent of all the votes were not for chocolate or vanilla?

_____

19. On his last 5 math quizzes, Paulo got the following scores: 97%, $\frac{3}{4}$, 82%, $\frac{91}{100}$, and $\frac{7}{10}$. What was his average quiz score?

_____

Holt Mathematics

Name _____ Date _____ Class _____

LESSON 7-7 Reteach
Percents

A percent is a ratio of a number to 100. Percent means "per hundred."

To write 38% as a fraction, write a fraction with a denominator of 100.

$$\frac{38}{100}$$

Then write the fraction in simplest form.

$$\frac{38}{100} = \frac{38 \div 2}{100 \div 2} = \frac{19}{50}$$

So, $38\% = \frac{19}{50}$.

Write each percent as a fraction in simplest form.

1. 43%   2. 72%   3. 88%   4. 35%

_____ _____ _____ _____

To write 38% as a decimal, first write it as fraction.

$$38\% = \frac{38}{100}$$

$\frac{38}{100}$ means "38 divided by 100."

```
        0.38
   100)38.00
       -300
        800
       -800
          0
```

So, 38% = 0.38.

Write each percent as a decimal.

5. 64%   6. 92%   7. 73%   8. 33%

_____ _____ _____ _____

Name _____ Date _____ Class _____

LESSON 7-7 Challenge
Per State

To show a percent, you can shade a 10-by-10 grid in any design that you want. For each percent below, try to shade the grid to look like the state it describes.

1. California has the largest population of any state. About 12% of all Americans live in California.

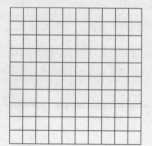

2. Florida is the top tourist state. About 26% of all visitors to the United States choose Florida for their vacations.

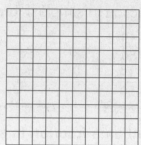

3. Nevada is the fastest-growing state. Its population has grown about 66% in the last ten years.

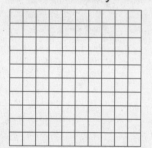

4. Alaska is the largest state. It makes up about 15% of the total area of the United States.

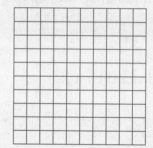

5. Washington produces the most apples. About 50% of all the apples grown in the U.S. come from Washington.

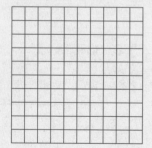

6. Texas is the top oil-producing state. About 21% of all the oil produced in the United States comes from Texas.

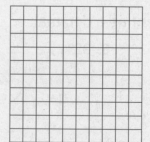

Copyright © by Holt, Rinehart and Winston.
All rights reserved.

Holt Mathematics

Name _____ Date _____ Class _____

LESSON 7-7 Problem Solving
Percents

Use the circle graph to answer each question. Write fractions in simplest form.

1. What fraction of the total 2000 music sales in the United States were rock recordings?

   _____

2. On this grid, model the percent of total United States music sales that were rap recordings. Then write that percent as a decimal.

   _____

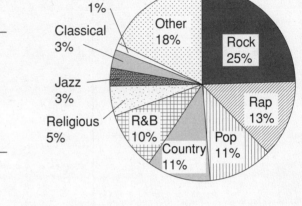

U.S. Recorded Music Sales, 2000

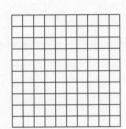

Circle the letter of the correct answer.

3. What kind of music made up $\frac{1}{20}$ of the total U.S. music recording sales?

   A Oldie   C Jazz
   B Classical   D Religious

4. What fraction of the United States music sales were country recordings?

   F $\frac{110}{100}$ H $\frac{1}{10}$
   G $\frac{11}{100}$ J $\frac{1}{100}$

5. What fraction of all United States recording sales did jazz and classical music make up together?

   A $\frac{6}{10}$ C $\frac{1}{5}$
   B $\frac{3}{50}$ D $\frac{11}{100}$

6. What kind of music made up $\frac{1}{10}$ of the total music recording sales in the United States in 2000?

   F Pop   H R&B
   G Jazz   J Oldies

Name _____ Date _____ Class _____

LESSON 7-7 Reading Strategies
Use Graphic Aids

The word **percent** means "per hundred." It is a ratio that compares a number to 100. A grid with 100 squares is used to picture percents.

Twelve percent is pictured on the grid below.

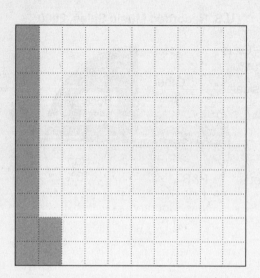

12 percent is a ratio, and means 12 per hundred. → $\frac{12}{100}$

12 percent can be written with symbols. → 12%

Use this figure to complete Exercises 1–4.

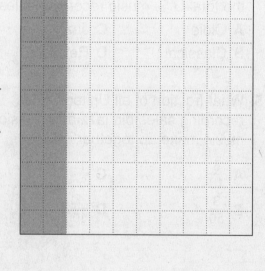

1. What is the ratio of shaded squares to the total number of squares? _____

2. Write the shaded amount using the % symbol.
_____

3. What is the ratio of unshaded squares to total number of squares? _____

4. Use the % symbol to write the unshaded amount. _____

Copyright © by Holt, Rinehart and Winston.
All rights reserved.

Holt Mathematics

Puzzles, Twisters & Teasers
7-7 Perfect 100!

Decide whether each statement is true or false. Circle your answer.

Unscramble your circled letters to spell an important word to know.

1. $45\% < \frac{1}{2}$ **N** true   **B** false
2. $6.4 = 64\%$ **A** true   **R** false
3. A 7% tax rate means you will pay $7 tax on a $10 purchase.

   W true   **T** false
4. $\frac{1}{5} = 20\%$ **P** true   **H** false
5. Thirty-seven percent is greater than one-third.

   E true   **I** false
6. Three-quarters is more than 80%.   **K** true   **C** false
7. $22\% = \frac{2}{11}$ **M** true   **E** false

Answer: _____

Name _____ Date _____ Class _____

LESSON 7-8
Practice A
Percents, Decimals, and Fractions

Write each decimal as a percent.

1. 0.1
_____

2. 0.6
_____

3. 0.02
_____

4. 0.14
_____

5. 0.22
_____

6. 0.03
_____

7. 0.25
_____

8. 0.17
_____

9. 0.39
_____

10. 0.8
_____

11. 0.04
_____

12. 0.99
_____

Write each fraction as a percent.

13. $\frac{1}{2}$
_____

14. $\frac{1}{4}$
_____

15. $\frac{3}{4}$
_____

16. $\frac{7}{10}$
_____

17. $\frac{97}{100}$
_____

18. $\frac{33}{100}$
_____

19. Brett scored $\frac{1}{4}$ of all the baskets he shot during the basketball game. What percent did he make?

_____

20. Sarah has 3 dimes and 1 nickel. Jamie has 2 quarters. What percent of a dollar do they each have?

_____

21. Mike, Joey, and Kini are playing a shooting game at the fair. Mike made $\frac{3}{5}$ of his shots, Joey made $\frac{4}{5}$, and Kini made $\frac{2}{5}$. Write the percent each boy made.

_____

Copyright © by Holt, Rinehart and Winston.
All rights reserved.

Holt Mathematics

Name _____ Date _____ Class _____

LESSON 7-8 Practice B
Percents, Decimals, and Fractions

Write each decimal as a percent.

1. 0.03

2. 0.92

3. 0.18

4. 0.49

5. 0.7

6. 0.09

7. 0.26

8. 0.11

9. 1.0

Write each fraction as a percent.

10. $\frac{2}{5}$

11. $\frac{1}{5}$

12. $\frac{7}{10}$

13. $\frac{1}{20}$

14. $\frac{1}{50}$

15. $\frac{4}{50}$

Compare. Write <, >, or = .

16. 60% ☐ $\frac{2}{3}$

17. 0.4 ☐ $\frac{2}{5}$

18. 0.5 ☐ 5%

19. $\frac{1}{100}$ ☐ 0.03

20. $\frac{7}{9}$ ☐ 72%

21. $\frac{3}{10}$ ☐ 35%

22. Bradley completed $\frac{3}{5}$ of his homework. What percent of his homework does he still need to complete?

23. After reading a book for English class, 100 students were asked whether or not they enjoyed it. Nine twenty-fifths of the students did not like the book. How many students liked the book?

Name _____ Date _____ Class _____

LESSON 7-8 Practice C
Percents, Decimals, and Fractions

Write each decimal as a percent and as a fraction or mixed number.

1. 0.96 _____
2. 0.04 _____
3. 0.28 _____

4. 0.65 _____
5. 0.32 _____
6. 0.005 _____

7. 1.13 _____
8. 2.08 _____
9. 3.002 _____

Write each fraction as a percent and as a decimal. Round to the nearest hundredth if necessary.

10. $\frac{12}{13}$ _____
11. $\frac{22}{27}$ _____
12. $\frac{15}{26}$ _____

13. $\frac{9}{31}$ _____
14. $\frac{34}{35}$ _____
15. $\frac{11}{23}$ _____

Compare. Write <, >, or =.

16. $\frac{12}{17}$ ☐ 77%
17. 0.18 ☐ $\frac{11}{25}$
18. $\frac{11}{50}$ ☐ 0.22
19. $\frac{21}{33}$ ☐ 80%
20. 0.4 ☐ $\frac{9}{32}$
21. $\frac{5}{16}$ ☐ 28%

22. During a sale, everything in the store was $\frac{1}{5}$ off the ticketed price. What percent of an item's original price should you expect to pay?

_____

23. Your teacher has offered you a choice for your 50 homework problems. You can do 48% of the problems, all of the even-numbered problems, or $\frac{3}{5}$ of the problems. Which option will you choose? How many problems will you have to do for homework?

_____

_____

Name _____ Date _____ Class _____

LESSON 7-8 Reteach
Percents, Decimals, and Fractions

You can write decimals as percents.

To write 0.5 as a percent, multiply the decimal by 100%.

0.5 • 100% = 50%

To multiply a number by 100, move the decimal point two places to the right.

0.50

So, 0.5 = 50%.

Write each decimal as a percent.

1. 0.8
2. 0.64
3. 0.075
4. 0.29

You can solve a proportion to write a fraction as a percent.

To write $\frac{3}{4}$ as a percent, first set up a proportion.

$\frac{3}{4} = \frac{x}{100}$

$3 \cdot 100 = 4 \cdot x$ The cross products are equal.

$300 = 4x$ x is multiplied by 4.

$\frac{4x}{4} = \frac{300}{4}$ Divide both sides by 4.

$x = 75$

So, $\frac{3}{4} = \frac{75}{100}$

$\frac{75}{100} = 75\%$, So, $\frac{3}{4} = 75\%$.

Write each fraction as a percent.

5. $\frac{4}{5}$
6. $\frac{9}{10}$
7. $\frac{1}{8}$
8. $\frac{7}{25}$

9. $\frac{1}{4}$
10. $\frac{5}{6}$
11. $\frac{3}{4}$
12. $\frac{1}{5}$

Challenge
LESSON 7-8
Trash or Treasure?

People in the United States produce about 208 million tons of garbage every year! We recycle about 56 million tons of that garbage, or about 27% of the total.

Complete the chart at right. Then display the percents on the circle graph below. Remember to give your graph a title. Label each section of the graph with the material and the percent of the total garbage recycled that each section represents. You may wish to color each section differently or add illustrations.

United States Recycling

Material	Total Garbage Recycled	
	Fraction	Percent
Metals	$\frac{1}{10}$	
Yard Waste	$\frac{17}{100}$	
Glass	$\frac{3}{50}$	
Paper	$\frac{29}{50}$	
Plastics	$\frac{1}{50}$	
All Other Materials	$\frac{7}{100}$	

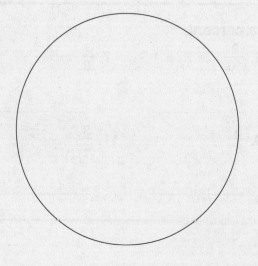

Name _____ Date _____ Class _____

LESSON 7-8 Problem Solving
Percents, Decimals, and Fractions

Write the correct answer.

1. Deserts cover about $\frac{1}{7}$ of all the land on Earth. About what percent of Earth's land is made up of deserts?

2. The Sahara is the largest desert in the world. It covers about 3% of the total area of Africa. What decimal expresses this percent?

3. Cactus plants survive in deserts by storing water in their thick stems. In fact, water makes up $\frac{3}{4}$ of the saguaro cactus's total weight. What percent of its weight is water?

4. Daytime temperatures in the Sahara can reach 130°F! At night, however, the temperature can drop by 62%. What decimal expresses this percent?

Circle the letter of the correct answer.

5. The desert nation of Saudi Arabia is the world's largest oil producer. About $\frac{1}{4}$ of all the oil imported to the United States is shipped from Saudi Arabia. What percent of our nation's oil is that?
   A 20%
   B 22%
   C 25%
   D 40%

6. About $\frac{2}{5}$ of all the food produced on Earth is grown on irrigated cropland. What percent of the world's food production relies on irrigation? What is the percent written as a decimal?
   F 40%; 40.0
   G 40%; 4.0
   H 40%; 0.4
   J 40%; 0.04

7. About $\frac{3}{25}$ of all the freshwater in the United States is used for drinking, washing, and other domestic purposes. What percent of our fresh water resources is that?
   A 3%
   B 25%
   C 12%
   D $\frac{1}{5}$

8. Factories and other industrial users account for about $\frac{23}{50}$ of the total water usage in the United States. Which of the following show that amount as a percent and decimal?
   F 46% and 0.46
   G 23% and 0.23
   H 50% and 0.5
   J 46% and 4.6

Name _____ Date _____ Class _____

LESSON 7-8 Reading Strategies
Multiple Meanings

A person can go by different names. Timothy could also be called Tim or Timmy.

A number can have different names too. The columns below show different names for 0.4. → Read: "four tenths."

Decimal Form **Fraction Form** **Percent Form**

0.4 → $\frac{4}{10}$ →   40%

To find the percent form of a fraction or decimal number, write an equivalent fraction with a denominator of 100.

$\frac{4}{10} = \frac{40}{100}$

A fraction with a denominator of 100 can be written as a percent. → $\frac{4}{10}$ → 40%

Use 0.37 to complete Exercises 1–3.

1. Write the words for 0.37. _____

2. Write 0.37 as a fraction. _____

3. Write 0.37 as a percent. _____

Use $\frac{60}{100}$ to complete Exercises 4–7.

4. How would you read $\frac{60}{100}$? _____

5. Write $\frac{60}{100}$ as a decimal. _____

6. Write $\frac{60}{100}$ as a percent. _____

Name _____ Date _____ Class _____

Puzzles, Twisters & Teasers
LESSON 7-8 *Chitter-Chatter!*

Some people talk all the time, while others never say a word. Have you ever wondered which animal talks the most? You are about to find out!

In the box below, cross out any number that is greater than one.

Next, cross out any pairs that have equivalent values. (For example, $\frac{1}{2}$ is the same as 50%).

Finally, order the remaining numbers from smallest to largest and place the corresponding letters in the answer spaces below.

$\frac{3}{4}$ **D** $\frac{1}{3}$ **Y**

                121% **P**   53% **A**

0.88 **K** $\frac{1}{4}$ **T**

            75% **V**

                                0.16 **A**

        0.25 **E**

2.06% **M** $\frac{3}{2}$ **R**

Place the remaining numbers (you should have 4 left) in order from smallest to largest in the spaces below.

What animal talks the most?

__A__ __Y__ __A__ __K__

Name _____ Date _____ Class _____

LESSON 7-9
Practice A
Percent Problems

Find the percent of each number.

1. 10% of 30 _____
2. 30% of 90 _____
3. 20% of 40 _____
4. 50% of 14 _____
5. 2% of 10 _____
6. 15% of 6 _____
7. 5% of 20 _____
8. 60% of 10 _____
9. 50% of 50 _____
10. 4% of 4 _____
11. 90% of 10 _____
12. 10% of 25 _____
13. 25% of 100 _____
14. 70% of 10 _____
15. 75% of 100 _____
16. 35% of 15 _____
17. 25% of 20 _____
18. 8% of 16 _____

19. Courtney made 12 model racecars. She painted 75% of her cars blue. How many of Courtney's racecar models are blue?

_____

20. Tim used 16 large beads to make a necklace. He chose bright orange for 25% of those beads. How many beads on Tim's necklace are bright orange?

_____

21. Taylor has 25 stuffed animals. She took 20% of those animals with her to a slumber party. How many stuffed animals did Taylor take to the slumber party?

_____

Holt Mathematics

Name _____ Date _____ Class _____

Practice B
LESSON 7-9 Percent Problems

Find the percent of each number.

1. 8% of 40 _____
2. 105% of 80 _____
3. 35% of 300 _____
4. 13% of 66 _____
5. 64% of 50 _____
6. 51% of 445 _____
7. 14% of 56 _____
8. 98% of 72 _____
9. 24% of 230 _____
10. 35% of 225 _____
11. 44% of 89 _____
12. 3% of 114 _____
13. 70% of 68 _____
14. 1.5% of 300 _____
15. 85% of 240 _____
16. 47% of 13 _____
17. 20% of 522 _____
18. 2.5% of 400 _____

19. Jenna ordered 28 shirts for her soccer team. Seventy-five percent of those shirts were size large. How many large shirts did Jenna order?

20. Douglas sold 125 sandwiches to raise money for his boy scout troop. Eighty percent of those sandwiches were sold in his neighborhood. How many sandwiches did Douglas sell in his neighborhood?

21. Samuel has run for 45 minutes. If he has completed 60% of his run, how many minutes will Samuel run in all?

Name _____ Date _____ Class _____

LESSON 7-9 Practice C
Percent Problems

Find the percent of each number.

1. 22% of 22 _____
2. 147% of 600 _____
3. 16% of 48 _____
4. 65% of 1,185 _____
5. 96% of 12 _____
6. 9% of 29 _____
7. 25% of 455 _____
8. 77% of 326 _____
9. 87% of 113 _____
10. 15.6% of 470 _____
11. 92% of 514 _____
12. 2.5% of 16 _____
13. 7.2% of 65 _____
14. 84.2% of 65 _____
15. 4.5% of 880 _____
16. 36.8% of 400 _____
17. 6.5% of 250 _____
18. 211% of 22 _____

19. The soccer team ordered 140 T-shirts to sell at the school fair. Oh those T-shirts, 50% are white, 20% are blue, 15% are green, 10% are red, and 5% are black. How many black T-shirts did the soccer team order? how many red?

_____

20. The Johnsons ordered new carpet for their family room. They paid 33% of the total cost when they ordered it, and will pay the remaining amount when the carpet is delivered. If they paid $192.72 when they ordered the carpet, how much will they pay when it is delivered?

_____

21. The city is going to raise its sales tax from 6.25% to 8.5% after the first of the year. How much tax would someone save on a $19,540 car if they bought the car before the first of the year rather than after the first of the year?

_____

Name _____ Date _____ Class _____

LESSON 7-9 Reteach
Percent Problems

You can use proportions to solve percent problems.

To find 25% of 72, first set up a proportion.

$$\frac{25}{100} = \frac{x}{72}$$

$25 \cdot 72 = 100 \cdot x$ Next, find cross products.

$1{,}800 = 100x$

$$\frac{100x}{100} = \frac{1{,}800}{100}$$ Then solve the equation.

$x = 18$

So, 18 is 25% of 72.

Use a proportion to find each number.

1. Find 3% of 75.   2. Find 15% of 85.   3. Find 20% of 50.   4. Find 6% of 90.

_____   _____   _____   _____

You can use multiplication to solve percent problems.

To find 9% of 70, first write the percent as a decimal.
    $9\% = 0.09$

Then multiply using the decimal.
    $0.09 \cdot 70 = 6.3$

So, 9% of 70 = 6.3.

Use multiplication to find each number.

5. Find 80% of 48.   6. Find 6% of 30.   7. Find 40% of 120.   8. Find 20% of 98.

_____   _____   _____   _____

9. Find 70% of 70.   10. Find 35% of 120.   11. Find 9% of 50.   12. Find 40% of 150.

_____   _____   _____   _____

Name _____ Date _____ Class _____

LESSON 7-9 Challenge
Pet Percentages

The United States Census Bureau counts all the people in the United States—but they do not count our pets! So, veterinarians use the percents shown in the table below to estimate pet populations. Their estimated U.S. pet population data is based on the 2000 census, which counted about 106 million households in the United States.

U.S. Pet Census, 2000

Pet	Percent of all Households	Estimated U.S. Pet Population
Dogs	53%	56,180,000
Cats	60%	63,600,000
Birds	13%	13,780,000
Horses	4%	4,240,000

Use the percents to estimate the number of pets that your class owns altogether, and the number of pets that your school owns altogether. Let each student in your class and each student in your school represent 1 household.

My Class and School Pet Population

Pet	Estimated Class Pet Population	Estimated School Pet Population
Dogs		
Cats		
Birds		
Horses		

Name _____ Date _____ Class _____

LESSON 7-9 Problem Solving
Percent Problems

In 2000, the population of the United States was about 280 million people.
Use this information to answer each question.

1. About 20% of the total United States population is 14 years old or younger. How many people is that?

   _____

2. About 6% of the total United States population is 75 years old or older. How many people is that?

   _____

3. About 50% of Americans live in states that border the Atlantic or Pacific Ocean. How many people is that?

   _____

4. About 12% of all Americans live in California. What is the population of California?

   _____

5. About 7.5% of all Americans live in the New York City metropolitan area. What is the population of that region?

   _____

6. About 12.3% of all Americans have Hispanic ancestors. What is the Hispanic American population here?

   _____

Circle the letter of the correct answer.

7. Males make up about 49% of the total population of the United States. How many males live here?
   A 1,372 million   C 13.72 million
   B 137.2 million   D 1.372 million

8. About 75% of all Americans live in urban areas. How many Americans live in or near large cities?
   F 70 milliom   H 210 million
   G 200 million   J 420 million

9. About 7.4% of all Americans live in Texas. What is the population of Texas?
   A 74 million   C 7.4 million
   B 20.72 million   D 2.072 million

10. Between 1990 and 2000, the population of the United States grew by about 12%. What was the U.S. population in 1990?
    F 250 million   H 313.6 million
    G 33.6 million   J 268 million

Holt Mathematics

Name _____ Date _____ Class _____

LESSON 7-9 Reading Strategies
Connect Words and Symbols

There are 24 students in Mrs. Wilson's class. Twenty-five percent of them take the bus to school. How many students ride the bus to school?

Step 1: Write a statement for the problem. → 25% of 24 students take the bus.

Step 2: Write an equation for the problem. → 25% of 24 is what number?

Step 3: Use symbols in place of words. → 25% • 24 = x

Step 4: Change the percent to a decimal. → 0.25 • 24 = x

Step 5: Multiply to solve. → $x = 6$

Answer each question.

1. What symbol stands for "of"?
   _____

2. What does x stand for in this problem?
   _____

3. Write the decimal value for 25%.
   _____

30% of the class brings their lunch to school. There are 50 sixth graders in the class. How many students bring their lunch?

4. Write a statement for this problem.
   _____

5. Rewrite the statement, using symbols for "of" and "is".
   _____

6. Rewrite the problem using a decimal in place of 30%.
   _____

7. Multiply to solve. How many students bring their lunch?
   _____

Name _____ Date _____ Class _____

LESSON 7-9 Puzzles, Twisters & Teasers
Odd Man Out!

In the problems below, two of the expressions mean the same thing, but one is different. Circle the one that is different. Write the circled letters in the corresponding spaces to solve the riddle.

1. 20% of 500 pages **A**   0.2 • 500 pages **W**   1,000 pages **P**

2. $\frac{40}{1.25}$ **L**   50 **D**   125% of 40 **T**

3. 20 minutes **R**   30% of an hour **N** $\frac{1}{3}$ of an hour **F**

4. 40% complete **S**   job is half-done **V**   60% more to do **I**

5. 20 ÷ 5 discount **O**   20% discount **K** $\frac{1}{5}$ reduction in price **M**

6. $\frac{3.6}{9}$ **B**   40% **H**   4 **E**

What starts with "E", ends with "E," but contains only one letter?

E __ __ __ __ __ __ E
   3 4 6 2 5 1

Name _____ Date _____ Class _____

LESSON 7-10 Practice A
Using Percents

Write the correct answer.

1. Glenn bought some candy that cost $5.00. If he had to pay a 5% sales tax, how much did he pay for his candy in all?

2. Nathan has ordered a pizza that costs $20.00. He wants to give the delivery person a 20% tip. How much should the tip be?

3. Jasmine wanted to buy a new pair of shoes that cost $30.00. When she went to the store, she found that the shoes were on sale for 10% off. How much did Jasmin pay for her shoes?

4. Marie ordered a root beer float at the ice cream shop. The float was $2.00, and she paid a 6% sales tax. How much did Marie pay for her root beer float in all?

5. Taylor has a coupon for 15% off any item in the toy store. The remote-control airplane he wants is $40.00. How much will the airplane cost if Taylor uses his coupon?

6. Victor went to the barber to get a haircut. The haircut cost $9.00. Victor gave the barber a 25% tip. How much did Victor spend at the barber shop altogether?

7. A CD is on sale for $10.00. The sales tax rate is 6%. How much will the total cost be for the CD?

8. A video game costs $25.00. The sales tax is 7%. How much will the total cost be for the game?

9. A bead store has a sign that reads "10% off the regular price." If Janice wants to buy beads that regularly cost $6.00, how much will she pay for them after the store's discount?

10. Julie gets a 20% discount on all of the items in the clothing store where she works. If she buys a shirt that regularly costs $45.00, how much money will she save with her employee discount?

Holt Mathematics

Name _____ Date _____ Class _____

Practice B
LESSON 7-10 Using Percents

Write the correct answer.

1. Carl and Rita ate breakfast at the local diner. Their bill came to $11.48. They gave their waitress a tip that was 25% of the bill. How much money did they give the waitress for her tip?

2. The school's goal for the charity fundraiser was $3,000. They exceeded the goal by 22%. How much money for charity did the school raise at the event?

3. Rob had a 15% off coupon for the sporting goods store. He bought a tennis racket that had a regular ticket price of $94.00. How much did Rob spend on the racket after using his coupon?

4. Lisa's family ordered sandwiches to be delivered. The total bill was $21.85. They gave the delivery person a tip that was 20% of the bill. How much did they tip the delivery person?

5. A portable CD player costs $118.26. The sales tax rate is 7%. About how much will it cost to buy the CD player?

6. Kathy bought two CDs that each cost $14.95. The sales tax rate was 5%. About how much did Kathy pay in all?

7. Tom bought $65.86 worth of books at the book fair. He got a 12% discount since he volunteered at the fair. About how much did Tom's books cost after the discount?

8. Sawyer bought a T-shirt for $12.78 and shorts for $17.97. The sales tax rate was 6%. About how much money did Sawyer spend altogether?

9. Melody buys a skateboard that costs $79.81 and a helmet that costs $26.41. She uses a 45% off coupon on the purchase. If Melody pays with a $100 bill, about how much change should she get back?

10. Bruce saved $35.00 to buy a new video game. The game's original price was $42.00, but it was on sale for 30% off. The sales tax rate was 5%. Did Bruce have enough money to buy the game? Explain.

Name _____ Date _____ Class _____

LESSON 7-10 Practice C
Using Percents

Write the correct answer.

1. A computer costs $979.99. The sales tax rate is 7%. How much will the total cost be for the computer?

2. Sheila bought $146.87 worth of groceries. The sales tax rate was 6%. How much did she spend in all?

3. Paul has saved $37.50 to buy a hamster, a cage, and hamster food. The hamster cost $5.50. The cage is on sale for 25% off the original price of $29.90. The food cost $2.64. The sales tax on the total is 5%. How much will Paul pay in all? How much of his savings will he have left over?

4. Jake has $65.50 to buy a new pair of jeans and a shirt. The jeans he wants cost $42.50, and the shirt costs $29.50. He has a coupon for 15% off, and the sales tax is 5%. Will he have enough money? Explain.

5. The bike Henry wants usually costs $147.99. Today, it is on sale for 15% off. After an 8% sales tax, how much will Henry pay for the bike?

6. Scott's lunch bill is $11.79. He gets an employee discount of 10% off. He leaves a 20% tip for the waitress. How much does Scott spend for lunch in all?

7. At Paint City, a gallon of paint with a regular price of $17.99 is now 15% off. At Giant Hardware, the same paint usually costs $21.99, but is now 24% off. Which store is offering the better deal?

8. Chelsea and Raymond's dinner bill was $57.82. They left the waitress a 26% tip. If they split the total cost of dinner evenly, how much did they each pay?

9. A store is having a going out of business sale for 55% off the ticketed prices. A pair of in-line skates has a ticketed price of $59.85, and a scooter has a ticketed price of $64.80. With a sales tax of 5%, how much will it cost to buy both items?

10. Troy works in a sporting goods store and gets an employee discount of 15% off any purchase. He wants to buy a fishing pole with a regular price of $70.00. The fishing pole is on sale for 33% off. How much will Troy pay for the fishing pole?

Copyright © by Holt, Rinehart and Winston.
All rights reserved.

78

Holt Mathematics

Name _____ Date _____ Class _____

LESSON 7-10 Reteach
Using Percents

There are many uses for percents.

Common Uses of Percents

Discounts	A **discount** is an amount that is subtracted from the regular price of an item. discount = regular price • discount rate
Tips	A **tip** is an amount added to a bill. tip = total bill • tip rate
Sales Tax	**Sales tax** is an amount added to the price of an item. sales tax = purchase price • sales tax rate

Rachel is buying a sweater that costs $42. The sales tax rate is 5%. About how much will the total cost of the sweater be?

You can use fractions to find the amount of sales tax.

First round $42 to $40.

Think: 5% is equal to $\frac{1}{20}$.

So, the amount of tax is about $\frac{1}{20}$ • $40.

The tax is about $2.00.

Then find the sum of the price of the sweater and the tax.

$42 + $2.00 = $44.00

Rachel will pay about $44.00 for the sweater.

Solve each problem.

1. About how much would you pay for a meal that costs $29.75 if you left a 15% tip?

2. About how much do you save if a book whose regular price is $25.00 is on sale for 10% off?

3. About how much would you pay for a box of markers whose price is $5.99 with a sales tax rate of 9.5%?

Name _____ Date _____ Class _____

LESSON 7-10

Challenge
Shop Smart

The Sport Zone and Sport City are both competing for customers by offering big discounts. To be a smart customer, you need to decide which store is offering the better price on each item.

For each item, write the store offering the best deal and the price you will pay there to the nearest whole cent.

The Sport Zone   Sport City

1.     The Sport Zone; $118.99

2.    Sport City; $30.30

3.     The Sport Zone; $15.50

4.     Sport City; $42.20

Name _____ Date _____ Class _____

LESSON 7-10 Problem Solving
Using Percents

Use the table to answer each question.

Federal Income Tax Rates, 2001

Single Income	Tax Rate	Married Joint Income	Tax Rate
$0 to $27,050	15%	$0 to $45,200	15%
$27,051 to $65,550	27.5%	$45,201 to $109,250	27.5%
$65,551 to $136,740	30.5%	$109,251 to $166,500	30.5%
$136,741 to $297,350	35.5%	$166,501 to $297,350	35.5%
More than $297,350	39.1%	More than $297,350	31.5%

1. If a single person makes $25,000 a year, how much federal income tax will he or she have to pay?

   _____

2. If a married couple makes $148,000 together, how much federal income tax will they have to pay?

   _____

3. The average salary for a public school teacher in the United States is $42,898. If two teachers are married, what is the average amount of federal income taxes they have to pay together?

   _____

4. In 2002 President George W. Bush received an annual salary of $400,000. Vice President Dick Cheney got $186,300. How much federal income tax do they each have to pay on their salary if they are married and filing jointly?

   _____

Circle the letter of the correct answer.

5. Members of the U.S. Congress each earn $145,100 a year. How much federal income tax does each pay on their salary?

   A $51,510.50   C $21,765
   B $44,255.50   D $39,902.50

6. A married couple each working a minimum-wage job will earn an average of $21,424 together a year. How much income tax will they pay?

   F $5,891.60   H $321.36
   G $3,213.60   J $6,534.32

7. The average American with a college degree earns $33,365 a year. About how much federal income tax does he or she have to pay at a single rate?

   A $5,004.75   C $10,176.33
   B $9,175.38   D $11,844.58

8. The governor of New York makes $179,000 a year. How much federal income tax does that governor have to pay at a single rate?

   F $63,545   H $49,225
   G $54,595   J $26,850

81 Holt Mathematics

Name _____ Date _____ Class _____

LESSON 7-10 Reading Strategies
Use a Graphic Organizer

This chart shows common ways percents are used. It also shows you how to figure a discount, sales price, sales tax, total price, tip, and total cost of a meal.

Discount	Sales Tax
An amount subtracted from the regular price of an item • Discount = regular price • discount rate • Sales price = regular price − the discount	An amount added to the price of an item • Sales tax = purchase price • sales tax rate • Total price = regular price + sales tax

Uses for Percents

Tip

The amount added to a bill for service
• Tip = price of meal • tip rate

Use the graphic organizer to answer each question.

1. What do you call the amount subtracted from the regular price of an item?

   _____

2. What do you call the amount added to a bill for service?

   _____

3. What do you call an amount added to the price of an item?

   _____

4. How do you find the discount?

   _____

5. How can you find the amount of sales tax on an item?

   _____

6. How is a total bill for a meal figured?

   _____

Name _____ Date _____ Class _____

Puzzles, Twisters & Teasers
LESSON 7-10 *Teacher's Favorite!*

Decide whether each statement is true or false. If the statement is true, follow the directions to navigate the maze. If the statement is false, ignore the directions and go to the next problem. Unscramble the letters that you land on to solve the riddle.

1. A tip is an amount added to a bill. Begin at start and move four spaces up.

2. A discount is an amount added to a bill. Move five spaces diagonally down and to the left.

3. Sihla is buying several CDs, totaling $45. If the sales tax is 5%, she will pay $47.25 total. Move three spaces left. Then move three spaces diagonally down and to the left.

4. You can find 10% of a number by moving the decimal point one place to the right. Move 6 spaces up.

5. A sign in a store reads "15% off all items". This is the same as a 15 percent discount on all items. Move three spaces diagonally down and to the right. Then move to the right as far as you can.

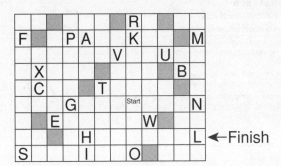

What is your teacher's favorite candy?

___ ___ ___ ___ ___ -- O L A T E

Practice A
7-1 Ratios and Rates

Use the table to write each ratio.

1. angel fish to tiger barbs 4:5
2. red-tail sharks to clown loaches 1:3
3. catfish to angel fish 1:4
4. clown loaches to tiger barbs 3:5
5. catfish to red-tail sharks 1:1

Caroline's Pet Fish	
Tiger Barbs	5
Catfish	1
Angel fish	4
Red-tail sharks	1
Clown loaches	3

6. Write three equivalent ratios to compare the number of black triangles in the picture with the total number of triangles.

Possible answer: 2:6, 1:3, 4:12

Use the table to write each ratio.

7. gray male kittens to gray female kittens
   2:5
8. white female kittens to white male kittens
   5:3

Caroline's Kittens		
	White	Gray
Male	3	2
Female	5	5

9. A candy store sells 2 ounces of chocolate for $0.80 and 3 ounces of chocolate for $0.90. How much does the store charge per ounce for the 2 ounces of chocolate? How much does the store charge per ounce for the 3 ounces of chocolate? Which is the better deal?

$0.40; $0.30; the 3 ounces of chocolate

Practice B
7-1 Ratios and Rates

Use the table to write each ratio.

1. lions to elephants 9:12 or 3:4
2. giraffes to otters 8:16 or 1:2
3. lions to seals 9:10
4. seals to elephants 10:12 or 5:6
5. elephants to lions 12:9 or 4:3

Animals in the Zoo	
Elephants	12
Giraffes	8
Lions	9
Seals	10
Otters	16

6. Write three equivalent ratios to compare the number of diamonds with the number of spades in the box.

Possible answer: 6:9, 2:3, 12:18

Use the table to write each ratio as a fraction.

7. Titans wins to Titans losses $\frac{12}{14}$ or $\frac{6}{7}$
8. Orioles losses to Orioles wins $\frac{15}{9}$ or $\frac{5}{3}$
9. Titans losses to Orioles losses $\frac{14}{15}$
10. Orioles wins to Titans wins $\frac{9}{12}$ or $\frac{3}{4}$

Baseball Team Stats		
	Titans	Orioles
Wins	12	9
Losses	14	15

11. A 6-ounce bag of raisins costs $2.46. An 8-ounce bag of raisins costs $3.20. Which is the better deal? the 8-ounce bag

12. Barry earns $36.00 for 6 hours of yard work. Henry earns $24.00 for 3 hours of yard work. Who has the better hourly rate of pay? Henry

---

Practice C
7-1 Ratios and Rates

Use the table to write each ratio.

1. red and blue T-shirts to green T-shirts
   66:36 or 7:6
2. purple T-shirts to yellow and green T-shirts
   51:96 or 17:32
3. blue and green T-shirts to purple and red T-shirts
   78:75 or 26:25
4. red T-shirts to all other T-shirt colors
   24:189 or 8:63

Store T-shirt Inventory, by Color	
Red	24
Blue	42
Green	36
Purple	51
Yellow	60

Write each ratio three different ways.

5. seven to twenty-one
   $\frac{7}{21}$; 7:21; 7 to 21
6. $\frac{12}{50}$
   12:50; 12 to 50; twelve to fifty
7. 18 to 10
   $\frac{18}{10}$; 18:10; eighteen to ten

Write three equivalent ratios for each ratio. Possible answers are given.

8. 19 to 38
   $\frac{1}{2}, \frac{2}{4}, \frac{3}{6}$
9. five to three
   $\frac{10}{6}, \frac{15}{9}, \frac{20}{12}$
10. $\frac{20}{24}$
    $\frac{10}{12}, \frac{5}{6}, \frac{15}{18}$

11. A 12-ounce bag of birdseed costs $3.12. A 16-ounce bag of birdseed costs $3.84. Which is the better deal? How much money per ounce would you save by buying that size bag instead of the other?

The 16-ounce bag; I would save $0.02 per ounce.

12. There are 60 players on a high school football team. The ratio of juniors and seniors to freshmen and sophomores on the team is 2:3. The ratio of juniors to seniors on the team is 1:2. How many juniors are on the team? How many seniors?

There are 8 juniors and 16 seniors on the team.

Reteach
7-1 Ratios and Rates

A ratio is a comparison of two quantities by division.

To compare the number of times vowels are used to the number of time consonants are used in the word "mathematics," first find each quantity.

Number of times vowels are used: 4
Number of times consonants are used: 7

Then write the comparison as a ratio, using the quantities in the same order as they appear in the word expression. There are three ways to write a ratio.

$\frac{4}{7}$ 4 to 7   4:7

Write each ratio.

1. days in May to days in a year
   31 to 365
2. sides of triangle to sides of a square
   3 to 4

Equivalent ratios are ratios that name the same comparison. The ratio of inches in a foot to inches in a yard is $\frac{12}{36}$. To find equivalent ratios, divide or multiply the numerator and denominator by the same number.

$\frac{12}{36} = \frac{12 \div 3}{36 \div 3} = \frac{4}{12}$ $\frac{12}{36} = \frac{12 \cdot 2}{36 \cdot 2} = \frac{24}{72}$

So, $\frac{12}{36}, \frac{4}{12}$, and $\frac{24}{72}$ are equivalent ratios.

Write three equivalent ratios to compare each of the following. Possible answers are given.

3. 8 triangles to 12 circles
   2:3, 4:6, 6:9
4. 20 pencils to 25 erasers
   4:5, 8:10, 12:15
5. 5 girls to 6 boys
   10:12, 15:18, 20:24
6. 10 pants to 14 shirts
   5:7, 15:21, 20:28

Holt Mathematics
84

Reteach
7-1 Ratios and Rates (continued)

A rate is a comparison of two quantities that have different units of measure.

Suppose a bus travels 150 miles in 3 hours. The rate could be written as $\frac{150 \text{ miles}}{3 \text{ hours}}$.

When the second term of a rate is 1 unit, the rate is a unit rate.

To write $\frac{150 \text{ miles}}{3 \text{ hours}}$ hours as a unit rate, divide each term by 3.

$\frac{150 \text{ miles}}{3 \text{ hours}}$
$= \frac{150 \text{ miles} \div 3}{3 \text{ hours} \div 3}$
$= \frac{50 \text{ miles}}{1 \text{ hour}}$

The unit rate is $\frac{50 \text{ miles}}{\text{hour}}$.

Find each unit rate.

7. $\frac{40 \text{ books}}{2 \text{ shelves}}$
   $\frac{20 \text{ books}}{\text{shelf}}$

8. $\frac{36 \text{ students}}{6 \text{ groups}}$
   $\frac{6 \text{ students}}{\text{group}}$

9. $\frac{300 \text{ seconds}}{5 \text{ minutes}}$
   $\frac{60 \text{ seconds}}{\text{min}}$

10. $\frac{54 \text{ miles}}{2 \text{ gallons}}$
    $\frac{27 \text{ miles}}{\text{gallon}}$

11. $\frac{4 \text{ miles}}{20 \text{ minutes}}$
    $\frac{0.2 \text{ miles}}{\text{min}}$

12. $\frac{\$1.29}{3 \text{ pounds}}$
    $\frac{\$0.43}{\text{pound}}$

13. $\frac{72 \text{ hours}}{3 \text{ days}}$
    $\frac{24 \text{ hours}}{\text{day}}$

14. $\frac{42 \text{ trading cards}}{6 \text{ packs}}$
    $\frac{7 \text{ trading cards}}{\text{pack}}$

Challenge
7-1 The Golden Ratio

For centuries, people all over the world have considered a certain rectangle to be one of the most beautiful shapes. Which of these rectangles do you find the most attractive?

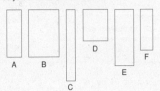

If you are like most people, you chose rectangle B. Why? It's a golden rectangle, of course! In a golden rectangle, the ratio of the length to the width is called the **golden ratio**—about 1.6 to 1.

Golden Ratio

 $w = 1$ in.

$\ell = 1.6$ in.

The golden ratio pops up all over the place—in music, sculptures, the Egyptian pyramids, seashells, paintings, pinecones, and of course in rectangles.

To create your own golden rectangle, just write a ratio equivalent to the golden ratio. This will give you the length and width of another golden rectangle.

Use a ruler to draw a new golden rectangle in the space below. Then draw several non-golden rectangles around it. Now conduct a survey of your family and friends to see if they choose the golden rectangle as their favorite.

Problem Solving
7-1 Ratios and Rates

Use the table to answer each question.

Atomic Particles of Elements

Element	Protons	Neutrons	Electrons
Gold	79	118	79
Iron	26	30	26
Neon	10	10	10
Platinum	78	117	78
Silver	47	61	47
Tin	50	69	50

1. What is the ratio of gold protons to silver protons?
   79:47

2. What is the ratio of gold neutrons to platinum protons?
   118:78 or 59:39

3. What are two equivalent ratios of the ratio of neon protons to tin protons?
   Possible answer: 10:50 and 1:5

4. What are two equivalent ratios of the ratio of iron protons to iron neutrons?
   Possible answer: 26:30 and 13:15

Circle the letter of the correct answer.

5. A ratio of one element's neutrons to another element's electrons is equivalent to 3 to 5. What are those two elements?
   (A) iron neutrons to tin electrons
   B gold neutrons to tin electrons
   C tin neutrons to gold electrons
   D neon neutrons to iron electrons

6. The ratio of two elements' protons is equivalent to 3 to 1. What are those two elements?
   F gold to tin
   G neon to tin
   (H) platinum to iron
   J silver to gold

7. Which element in the table has a ratio of 1 to 1, no matter what parts you are comparing in the ratio?
   A iron   **C** tin
   (B) neon   **D** silver

8. If the ratio for any element is 1:1, which two parts is the ratio comparing?
   F protons to neutrons
   G electrons to neutrons
   (H) protons to electrons
   J neutrons to electrons

Reading Strategies
7-1 Use the Context

A **ratio** is a comparison between two similar quantities. The picture below shows geometric figures. You can write ratios to compare the figures.

Compare the number of triangles to the total number of figures. This comparison can be written as a ratio in three different ways.

$\frac{\text{number of triangles}}{\text{total figures}}$ → $\frac{2}{9}$ Read: "two to nine."

2 to 9

2:9   Read: "two to nine."

Compare the number of squares to the number of circles.

1. Write the ratio that compares the number of squares to the number of circles in three different ways.
   $\frac{3}{4}$; 3 to 4; 3:4

A **rate** compares two different kinds of quantities. Rates can be shown in different ways.

You can buy 3 cans of juice for $4. The comparison of juice to money can be written:

$\frac{3 \text{ cans}}{\$4}$ → $\frac{3}{4}$ 3 to 4   3:4

Julie can jog eight miles in two hours. Use this information to complete Exercises 2–4.

2. Write the rate using words.   **eight miles in two hours**

3. Write the rate with numbers in three different ways.
   $\frac{8}{2}$, 8 to 2, 8:2

4. Compare ratios and rates. How are they alike?
   both compare two quantities

LESSON 7-1 Puzzles, Twisters & Teasers
Cool Runner!

Fill in the crossword puzzle. Unscramble the circled letters to solve the riddle.

Down
1. A comparison to 1 of something is a _____ rate.
2 & 4. Hector earns $24 for 3 hours of work, while Myrna earns $17 for 2 hours. We would say Myrna's rate is a __2__ __4__
3. A comparison of two quantities using division.
6. The ratio of boys to girls is 5:1. Number of girls if there are 10 boys.

Across
5. _____ ratios are different ratios that name the same comparison.
7. The number of boys when the ratio of boys to girls is 10:3 and the number of girls is 3.
8. The middle word when you read 10:7 ("ten _____ seven").
9. A _____ compares two quantities that have different units of measure.

What runs but never gets out of breath? W A T E R

Crossword answers:
- UNIT
- BETTER
- RATIO
- DIVIDED
- EQUIVALENT
- TWO
- TEN
- TO
- RATE

LESSON 7-2 Practice A
Using Tables to Explore Equivalent Ratios and Rates

Use each table to find three equivalent ratios. **Possible answers are given.**

1. $\frac{1}{5}$

1	2	3	4
5	10	15	20

Equivalent ratios: $\frac{1}{5}$, $\frac{2}{10}$, $\frac{3}{15}$, and $\frac{4}{20}$

2. 3 to 8

3	6	9	12
8	16	24	32

Equivalent ratios: 3 to 8, __6 to 16__, __9 to 24__, and __12 to 32__

3. 80:40

80	40	20	10
40	20	10	5

Equivalent ratios: 80:40, __40:20__, __20:10__, and __10:5__

Use a table to find three equivalent ratios. **Possible answers are given**

4. 3 to 6
   __6 to 12; 9 to 18; 12 to 24__

5. $\frac{7}{3}$
   $\frac{14}{6}, \frac{21}{9}, \frac{28}{12}$

6. 1:2
   __2:4; 3:6; 4:8__

7. 2 to 1
   __4 to 2; 6 to 3; 8 to 4__

8. Alan swims laps in a pool. The table shows how long it takes him to swim different numbers of laps.

Number of Laps	3	6	9	12	15
Time (min)	6	12	18	24	30

How long do you predict it will take Alan to swim 10 laps?
__20 minutes__

LESSON 7-2 Practice B
Using Tables to Explore Equivalent Ratios and Rates

Use a table to find three equivalent ratios. **Possible answers are given.**

1. 4 to 7
   __8 to 14; 12 to 21; 16 to 28__

2. $\frac{10}{3}$
   $\frac{20}{6}, \frac{30}{9}, \frac{40}{12}$

3. 2:5
   __4:10; 6:15; 8:20__

4. 8 to 9
   __16 to 18; 24 to 27; 32 to 36__

5. 3 to 15
   __6 to 30; 9 to 45; 12 to 60__

6. $\frac{30}{90}$
   $\frac{15}{45}, \frac{10}{30}, \frac{6}{18}$

7. 1:3
   __2:6; 3:9; 4:12__

8. $\frac{7}{2}$
   $\frac{14}{4}, \frac{21}{6}, \frac{28}{8}$

9. Britney does sit-ups every day. The table shows how long it takes her to do different numbers of sit-ups.

Number of Sit-Ups	10	30	50	200	220
Time (min)	2	6	10	40	44

How long do you predict it will take Britney to do 120 sit-ups?
__24 minutes__

10. The School Supply Store has markers on sale. The table shows some sale prices.

Number of Markers	12	8	6	4	2
Cost ($)	9.00	6.00	4.50	3.00	1.50

How much do you predict you would pay for 10 markers?
__$7.50__

LESSON 7-2 Practice C
Using Tables to Explore Equivalent Ratios and Rates

Use a table to find three equivalent ratios. **Possible answers are given.**

1. 5 to 11
   __10 to 22; 15 to 33; 20 to 44__

2. $\frac{17}{19}$
   $\frac{34}{38}, \frac{51}{57}, \frac{68}{76}$

3. 8:7
   __16:14; 24:21; 32:28__

4. 6 to 13
   __12 to 26; 18 to 39; 24 to 52__

5. 36 to 12
   __18 to 6; 12 to 4; 9 to 3__

6. $\frac{48}{90}$
   $\frac{24}{45}, \frac{16}{30}, \frac{8}{15}$

Multiply and divide each ratio to find two equivalent ratios. **Possible answers are given.**

7. 10:20
   __20:40; 5:10__

8. $\frac{6}{9}$
   $\frac{12}{18}, \frac{2}{3}$

9. $\frac{118}{66}$
   $\frac{236}{132}, \frac{59}{33}$

10. 25:100
    __50:200; 5:20__

11. Spring Street Middle School orders 9 calculators for every 12 students. The table shows how many calculators the school orders for certain numbers of students.

Students	12	36	60	96	240
Calculators	9	27	45	72	180

How many calculators do you predict the school would order for 132 students?
__99 calculators__

LESSON 7-2 Reteach
Using Tables to Explore Equivalent Ratios and Rates

You can use a table to find ratios equivalent to $\frac{1}{4}$.
Write the numerator in the top box for the original ratio.
Write the denominator in the bottom box for the original ratio.
Then multiply the numerator and the denominator by 2, 3, and 4.

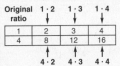

Use each new numerator and denominator to write an equivalent ratio.

So, the ratios $\frac{2}{8}$, $\frac{3}{12}$, and $\frac{4}{16}$ are equivalent to $\frac{1}{4}$.

1. Use the table to find three ratios equivalent to $\frac{2}{5}$. Possible answers are given.

2	4	6	8
5	10	15	20

Equivalent ratios: $\frac{2}{5}$, $\frac{4}{10}$, $\frac{6}{15}$, and $\frac{8}{20}$

You can use a table to find ratios equivalent to 3 to 8.
Write the first number in the top box for the original ratio.
Write the second number in the bottom box for the original ratio.
Then multiply both numbers by 2, 3, and 4.

Original ratio 3·2 3·3 3·4

3	6	9	12
8	16	24	32

Use each new top number and bottom number to write an equivalent ratio.

So, the ratios 6 to 16, 9 to 24, and 12 to 32 are equivalent to 3 to 8.

2. Use the table to find three ratios equivalent to 4 to 10. Possible answers are given.

4	8	12	16
10	20	30	40

Equivalent ratios: 4 to 10, 8 to 20, 12 to 30, and 16 to 40

LESSON 7-2 Challenge
It's All Black and White!

This grid has a black-to-white ratio of 5 to 4.

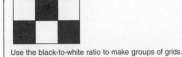

Use the black-to-white ratio to make groups of grids.
Then complete the table of equivalent ratios.

Black	5	10	15	20	25	30	35	40	45	50	55	60
White	4	8	12	16	20	24	28	32	36	40	44	48

LESSON 7-2 Problem Solving
Using Tables to Explore Equivalent Ratios and Rates

Use the table to answer the questions.

School Outing Student-to-Parent Ratios

Number of Students	8	16	24	32	40	48	56	64	72
Number of Parents	2	4	6	8	10	12	14	16	18

1. Each time some students go on a school outing, their teachers invite students' parents to accompany them. Predict how many parents will accompany 88 students.

   22 parents

2. Next week 112 students will go to the Science Museum. Their teachers invited some of the students' parents to go with them. How many parents do you predict will go with the students to the Science Museum?

   28 parents

Circle the letter of the correct answer.

3. Tanya's class of 28 students will be going to the Nature Center. How many parents do you predict Tanya's teacher will invite to accompany them?
   A 5 parents
   (B) 7 parents
   C 9 parents
   D 11 parents

4. Some students will be going on an outing to the local police station. Their teachers invited 13 parents to accompany them. How many students do you predict will be going on the outing?
   F 49 students
   G 50 students
   H 51 students
   (J) 52 students

5. In June, all of the students in the school will be going on their annual picnic. If there are 416 students in the school, what do you predict the number of parents accompanying them on the picnic will be?
   A 52 parents
   B 78 parents
   (C) 104 parents
   D 156 parents

6. On Tuesday, all of the sixth-grade students will be going to the Space Museum. Their teachers invited 21 parents to accompany them. How many sixth graders do you predict will be going to the Space Museum?
   F 80 sixth graders
   G 82 sixth graders
   (H) 84 sixth graders
   J 86 sixth graders

LESSON 7-2 Reading Strategies
Understand Vocabulary

Equivalent ratios are ratios that name the same comparison. The box below shows different ratios. You can find equivalent ratios by multiplying and dividing. Then you can organize them in a table.

| $\frac{3}{2}$ | 6 to 4 | $\frac{18}{10}$ | 15:10 | 12:8 |

Look for equivalent ratios. Start with $\frac{3}{2}$. Multiply the numerator and denominator by 2.
$\frac{3}{2} = \frac{3 \cdot 2}{2 \cdot 2} = \frac{6}{4}$
The resulting ratio is $\frac{6}{4}$. So 6 to 4 is equivalent to $\frac{3}{2}$.
Try $\frac{18}{10}$. Divide the numerator and denominator by 6.
$\frac{18}{10} = \frac{18 \div 6}{10 \div 6} = \frac{3}{1.7}$
The resulting ratio is not $\frac{3}{2}$. So $\frac{18}{10}$ is not equivalent to $\frac{3}{2}$.
Try 15:10. Divide each number by 5.
$15 \div 5 = 3$
$10 \div 5 = 2$
The resulting ratio is 3:2. So 15:10 is equivalent to $\frac{3}{2}$.
Try 12:8. Divide each number by 4.
$12 \div 4 = 3$
$8 \div 4 = 2$
The resulting ratio is 3:2. So 12:8 is equivalent to $\frac{3}{2}$.
Organize the equivalent ratios in a table. Write the ratios in order from least terms to greatest terms.

3	6	12	15
2	4	8	10

1. Find the equivalent ratios in the box.

| $\frac{25}{35}$ | 5 to 7 | 15:21 | 10 to 15 | $\frac{50}{70}$ |

Equivalent ratios: $\frac{25}{35}$, 5 to 7, 15:21, $\frac{50}{70}$

2. Organize the equivalent ratios in the table in order from least terms to greatest terms.

5	15	25	50
7	21	35	70

---

87 **Holt Mathematics**

LESSON 7-2 Puzzles, Twisters, and Teasers
Unlike the Others

Each row of problems has 3 equivalent ratios and 1 that is not. Circle the one that is not equivalent to the others. Write the circled letters in the corresponding spaces to solve the riddle.

1. $\frac{5}{10}$ **S**   20:30 (**N**)   10 to 20 **M** $\frac{1}{2}$ **P**

2. 24:36 **U** $\frac{12}{16}$ (**C**)   6:9 **Z**   48 to 72 **I**

3. $\frac{96}{120}$ **F**   24 to 30 **A**   12:15 **X** $\frac{5}{4}$ (**L**)

4. 88 to 100 **D**   22 to 25 **B**   116:200 (**E**) $\frac{264}{300}$ **O**

5. 9 to 5 **C** $\frac{81}{40}$ (**E**)   99:55 **K**   135 to 75 **Y**

No sooner spoken than broken. What is it?

S I __L__ __E__ __N__ __C__ __E__
       3   5   1   2   4

LESSON 7-3 Practice A
Proportions

Find the missing value in each proportion.

1. $\frac{1}{2} = \frac{n}{6}$
   $n = 3$

2. $\frac{6}{9} = \frac{n}{3}$
   $n = 2$

3. $\frac{n}{14} = \frac{2}{7}$
   $n = 4$

4. $\frac{2}{3} = \frac{6}{n}$
   $n = 9$

5. $\frac{n}{5} = \frac{12}{15}$
   $n = 4$

6. $\frac{2}{n} = \frac{1}{6}$
   $n = 12$

7. $\frac{10}{2} = \frac{n}{4}$
   $n = 20$

8. $\frac{1}{4} = \frac{2}{n}$
   $n = 8$

9. $\frac{16}{8} = \frac{n}{4}$
   $n = 8$

Write a proportion for each model.

10.

Possible answer: $\frac{4}{6} = \frac{2}{3}$

11.

Possible answer: $\frac{8}{10} = \frac{4}{5}$

12. Jeff made 2 out of every 5 baskets he shot during basketball practice. If he took 25 shots, how many baskets did he make?

10 baskets

13. Tyra gets 2 quarters for every 3 newspapers she delivers. If she delivers 21 newspapers, how many quarters will she get? How much money is that in all?

14 quarters; $3.50

LESSON 7-3 Practice B
Proportions

Find the missing value in each proportion.

1. $\frac{24}{8} = \frac{n}{2}$
   $n = 6$

2. $\frac{4}{9} = \frac{20}{n}$
   $n = 45$

3. $\frac{n}{36} = \frac{5}{6}$
   $n = 30$

4. $\frac{n}{5} = \frac{4}{10}$
   $n = 2$

5. $\frac{3}{9} = \frac{2}{n}$
   $n = 6$

6. $\frac{6}{n} = \frac{3}{7}$
   $n = 14$

7. $\frac{5}{3} = \frac{n}{6}$
   $n = 10$

8. $\frac{9}{6} = \frac{6}{n}$
   $n = 4$

9. $\frac{2}{130} = \frac{1}{n}$
   $n = 65$

Write a proportion for each model.

10.

Possible answer: $\frac{9}{12} = \frac{3}{4}$

11.

Possible answer: $\frac{16}{4} = \frac{4}{1}$

12. Shane's neighbor pledged $1.25 for every 0.5 miles that Shane swims in the charity swim-a-thon. If Shane swims 3 miles, how much money will his neighbor donate?

$7.50

13. Barbara's goal is to practice piano 20 minutes for every 5 minutes of lessons she takes. If she takes a 20 minute piano lesson this week, how many minutes should she practice this week?

80 minutes

LESSON 7-3 Practice C
Proportions

Find the missing value in each proportion.

1. $\frac{6}{15} = \frac{n}{45}$
   $n = 18$

2. $\frac{n}{160} = \frac{1}{40}$
   $n = 4$

3. $\frac{2}{8} = \frac{n}{56}$
   $n = 14$

4. $\frac{13}{26} = \frac{n}{4}$
   $n = 2$

5. $\frac{4}{9} = \frac{32}{n}$
   $n = 72$

6. $\frac{n}{16} = \frac{14}{32}$
   $n = 7$

7. $\frac{1}{17} = \frac{0.5}{n}$
   $n = 8.5$

8. $\frac{8.1}{9} = \frac{n}{15}$
   $n = 13.5$

9. $\frac{9.1}{7} = \frac{n}{5}$
   $n = 6.5$

10. Use circles and triangles to draw a model for the proportion $\frac{5}{6} = \frac{10}{12}$. Possible answer:

11. Use hearts and diamonds to draw a model for the proportion $\frac{3}{4} = \frac{9}{12}$. Possible answer:

12. To avoid dehydration, a person should drink 8 ounces of water for every 15 minutes of exercise. How much water should Hahn drink if he cycles for 135 minutes?

72 ounces

13. Leo has entered a reading contest to raise money for charity. His aunt has agreed to pay Leo $0.13 for every 5 pages that he reads. Leo's uncle has promised to match every whole dollar that Leo collects in the contest with $1.75. If Leo reads 365 pages, how much money will his aunt donate to the charity? How much will Leo's uncle give to match the aunt's donation?

$9.49; $15.75

LESSON 7-3 Reteach
Proportions

A proportion is an equation that shows two equivalent ratios.
$\frac{3}{4} = \frac{9}{12}$ is an example of a proportion.
$3 \cdot 12 = 36$ and $4 \cdot 9 = 36$.   The cross products of proportions are equal.

You can use cross products to find the missing value in a proportion.

$$\frac{3}{x} = \frac{12}{48}$$
$12 \cdot x = 3 \cdot 48$ To find x, first find the cross products.
$12x = 144$
Think: $144 \div 12 = x$ Then use a related math sentence to solve the equation.
$x = 12$
So, $\frac{3}{12} = \frac{12}{48}$.

Find the cross products to solve each proportion.

1. $\frac{x}{8} = \frac{3}{4}$
   $x \cdot 4 = 8 \cdot 3$
   $x = 6$

2. $\frac{2}{3} = \frac{x}{6}$
   $2 \cdot 6 = 3 \cdot x$
   $x = 4$

3. $\frac{2}{5} = \frac{4}{x}$
   $2 \cdot x = 5 \cdot 4$
   $x = 10$

4. $\frac{6}{x} = \frac{1}{3}$
   $6 \cdot 3 = x \cdot 1$
   $x = 18$

5. $\frac{3}{8} = \frac{12}{x}$
   $x = 32$

6. $\frac{3}{5} = \frac{6}{x}$
   $x = 10$

7. $\frac{x}{8} = \frac{2}{16}$
   $x = 1$

8. $\frac{2}{9} = \frac{4}{x}$
   $x = 18$

9. $\frac{3}{4} = \frac{15}{x}$
   $x = 20$

10. $\frac{1}{2} = \frac{x}{30}$
    $x = 15$

11. $\frac{x}{5} = \frac{24}{30}$
    $x = 4$

12. $\frac{25}{35} = \frac{5}{x}$
    $x = 7$

LESSON 7-3 Challenge
Patriotic Proportions

On August 21, 1959, President Eisenhower signed an order that established the official proportions of the United States flag. No matter what size the flag is, it must match those proportions to be used officially.

Official Proportions for the United States Flag	
Width of flag	1
Length of flag	$1\frac{9}{10}$
Width of union	$\frac{7}{13}$
Length of union	$\frac{19}{25}$
Width of each stripe	$\frac{1}{13}$

The union is the blue area. The 50 stars represent the 50 states.

The 13 stripes represent the first 13 states.

Use the official proportions to find the missing dimension of each flag.

1. Length of flag = 10 feet; Width of flag = $5\frac{5}{19}$ feet

2. Width of flag = 57 yards; Length of flag = $108\frac{3}{10}$ yards

3. Width of flag = 13 centimeters; Width of Union = 7 centimeters

4. Width of flag = 260 inches; Width of each stripe = 20 inches

5. Length of flag = 25 meters; Length of Union = 10 meters

Choose a width in inches for a United States flag. Then use a ruler to draw your flag with the official proportional length in the space below.

Check students' flag widths and lengths for the correct width-to-length ratio of 1 inch to 1.9 inches.

LESSON 7-3 Problem Solving
Proportions

Write the correct answer.

1. For most people, the ratio of the length of their head to their total height is 1:7. Use proportions to test your measurements and see if they match this ratio.

   Answers should test the 1:7 head to height ratio measurements.

2. The ratio of an object's weight on Earth to its weight on the Moon is 6:1. The first person to walk on the Moon was Neil Armstrong. He weighed 165 pounds on Earth. How much did he weigh on the Moon?

   27.5 pounds

3. It has been found that the distance from a person's eye to the end of the fingers of his outstretched hand is proportional to the distance between his eyes at a 10:1 ratio. If the distance between your eyes is 2.3 inches, what should the distance from your eye to your outstretched fingers be?

   23 inches

4. Chemists write the formula of ordinary sugar as $C_{12}H_{22}O_{11}$, which means that the ratios of 1 molecule of sugar are always 12 carbon atoms to 22 hydrogen atoms to 11 oxygen atoms. If there are 4 sugar molecules, how many atoms of each element will there be?

   48 carbon, 88 hydrogen, 44 oxygen

Circle the letter of the correct answer.

5. A healthy diet follows the ratio for meat to vegetables of 2.5 servings to 4 servings. If you eat 7 servings of meat a week, how many servings of vegetables should you eat?
   A 28 servings   C 14 servings
   B 17.5 servings (D) 11.2 servings

6. A 150-pound person will burn 100 calories while sitting still for 1 hour. Following this ratio, how many calories will a 100-pound person burn while sitting still for 1 hour?
   F $666\frac{2}{3}$ calories   H $6\frac{2}{3}$ calories
   (G) $66\frac{2}{3}$ calories   J 6 calories

7. Recently, 1 U.S. dollar was worth 1.58 in euros. If you exchanged $25 at that rate, how many euros would you get?
   (A) 39.50 euros
   B 15.82 euros
   C 26.58 euros
   D 23.42 euros

8. Recently, 1 U.S. dollar was worth 0.69 English pound. If you exchanged 500 English pounds, how many dollars would you get?
   F 345 U.S. dollars
   (G) 725 U.S. dollars
   H 500.69 U.S dollars
   J 499.31 U.S. dollars

LESSON 7-3 Reading Strategies
Use Graphic Aids

A **proportion** is a statement of two equal ratios. This statement is written as an equation.

One cup of juice contains 50 calories.

This statement can be written as a ratio.
$\frac{\text{cups}}{\text{calories}} \rightarrow \frac{1}{50}$

Two cups of juice contain 100 calories.

This statement can also be written as a ratio.
$\frac{\text{cups}}{\text{calories}} \rightarrow \frac{2}{100}$

Are these two ratios equal?

Step 1: Write a proportion with the two ratios.
$\frac{1}{50} = \frac{2}{100}$ → Read: "1 is to 50 as 2 is to 100."

Step 2: Find the cross products. If cross products are equal, the ratios are equal and form a proportion.
$\frac{1}{50} \times \frac{2}{100}$ $2 \times 50 = 100$
                                     $1 \times 100 = 100$

Use this picture to answer the questions.

1. What is the ratio of striped circles to total circles? $\frac{2}{8}$

2. What is the ratio of black circles to white circles? $\frac{1}{4}$

3. Find the cross products. Write = or ≠ to complete.
   $2 \times 4 = 8$
   $1 \times 8 = 8$
   $8 = 8$

4. Do $\frac{2}{8}$ and $\frac{1}{4}$ form a proportion?
   yes

LESSON 7-3 Puzzles, Twisters & Teasers
Too Much!

Solve the problems and circle your answers.

Using the letters next to your answers, create three words that mean different things, but are all pronounced the same! You will use two of the letters more than once.

1. Chung is giving medicine to his cat Princess. The bottle recommends 3 pills for a 15 pound cat, but Princess weighs only 10 pounds. How many pills should Chung give?

   R 1 pill   (W) 2 pills   A 3 pills

2. Find the missing value: $\frac{5}{n} = \frac{15}{21}$

   (O) 7   D 3   M 5

3. Suri knows that she needs to study about twenty minutes a night for each hour class in math, and about thirty minutes for each hour class in history. Normally she has one hour of each class every day. But today she had math class for an hour and a half and only a half-hour history class. Will her homework take more, less, or the same amount of time tonight?

   F more   P same   (T) less

Now use the letters next to your answers to figure out the three words.

A number: T W O
Also: T O O
A preposition: T O

LESSON 7-4 Practice A
Similar Figures

Tell whether the figures in each pair are similar.

1. similar   2. not similar
3. not similar   4. similar

5. The two triangles are similar. Find the missing length x and the measure of ∠F.

   $x = 4$ in.; $m\angle F = 50°$

6. The two triangles are similar. Find the missing length m and the measure of ∠O.

   $m = 7.4$ cm; $m\angle O = 60°$

7. Two rectangular photos are similar. The larger photo is 6 inches wide and 8 inches long. The smaller photo is 3 inches wide. What is the smaller photo's length?

   4 inches

8. Two triangular mirrors are similar. The first mirror's angles all measure 60°. What are the measures of the second mirror's angles? Explain how you know.

   They are all 60°; Corresponding angles in similar figures are congruent.

LESSON 7-4 Practice B
Similar Figures

Write the correct answers.

1. The two triangles are similar. Find the missing length x and the measure of ∠A.

   $x = 18$ ft; $m\angle A = 80°$

2. The two triangles are similar. Find the missing length x and the measure of ∠J.

   $x = 8$ m; $m\angle J = 23°$

3. The two triangles are similar. Find the missing length x and the measure of ∠N.

   $x = 24$ cm; $m\angle N = 53°$

4. Juanita planted two flower gardens in similar square shapes. What are the measures of all the angles in each garden? Explain how you know.

   They are all 90°; All squares have all right angles.

LESSON 7-4 Practice C
Similar Figures

The figures in each pair are similar. Find the unknown measures.

1. $w = 6.5$ in.

2. $m\angle A = 84°$, $m\angle X = 48°$, $m\angle Y = 48°$, $x = 12$ ft

3. $m\angle C = 70°$, $m\angle N = 110°$, $m\angle O = 70°$, $x = 4.2$ cm

4. $\angle C = 37°$, $\angle E = 53°$, $x = 12.5$ in.

5. Two regular pentagons are similar. One side of the first pentagon is 3 m long, and the perimeter of the second pentagon is three times as long as the first pentagon. What are the lengths of each side of the second pentagon?

   9 meters

6. A 7-by-9 foot rectangle is similar to a second rectangle whose perimeter is 260 ft. What are the dimensions of the second rectangle?

   56.875 ft by 73.125 ft

LESSON 7-4 Reteach
Similar Figures

Two figures are similar if they have the same shape but are different sizes.
Similar figures have corresponding sides and corresponding angles. Corresponding sides are proportional. Corresponding angles are congruent.

Look at the similar triangles below.

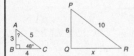

$\overline{AB}$ corresponds to $\overline{PQ}$.   ∠A corresponds to ∠P.
$\overline{BC}$ corresponds to $\overline{QR}$.   ∠B corresponds to ∠Q.
$\overline{AC}$ corresponds to $\overline{PR}$.   ∠C corresponds to ∠R.

What is the length of $\overline{QR}$?

$\frac{AB}{BC} = \frac{PQ}{QR}$ Set up a proportion.
$\frac{3}{4} = \frac{6}{x}$ Substitute the values.
$3 \cdot x = 4 \cdot 6$ The cross products are equal.
$3x = 24$ x is multiplied by 3.
$\frac{3x}{3} = \frac{24}{3}$ Divide both sides by 3.
$x = 8$

So, the length of $\overline{QR}$ is 8 units.

Find each missing length.

1.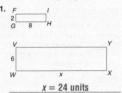

   x = 24 units

2.

   x = 26 units

LESSON 7-4 Challenge
You Won't Believe Your Eyes!

Answer each question by looking at the drawings below. Then use what you know about similar and congruent figures to verify your answers.

1. Are the two line segments congruent?

   yes

2.
   Are the two center circles similar or congruent?

   congruent

3. Are any of these circles similar?

   yes; all the circles

4. Are any of these line segments congruent?

   yes; horizontal lines

5. Which horizontal line is longer?

   Neither, they are congruent.

6. Which two figures are congruent? Which two figures are similar?

   congruent circles; similar squares

LESSON 7-4 Problem Solving
Similar Figures

Write the correct answer.

1. The map at right shows the dimensions of the Bermuda Triangle, a region of the Atlantic Ocean where many ships and airplanes have disappeared. If a theme park makes a swimming pool in a similar figure, and the longest side of the pool is 0.5 mile long, about how long would the other sides of the pool have to be?

   0.403 mile

2. Completed in 1883, *The Battle of Gettysburg* is 410 feet long and 70 feet tall. A museum shop sells a print of the painting that is similar to the original. The print is 2.05 feet long. How tall is the print?

   0.35 ft

3. *Panorama of the Mississippi* was 12 feet tall and 5,000 feet long! If you wanted to make a copy similar to the original that was 2 feet tall, how many feet long would the copy have to be?

   $833\frac{1}{3}$ feet

Circle the letter of the correct answer.

4. Two tables shaped like triangles are similar. The measure of one of the larger table's angles is 38°, and another angle is half that size. What are the measures of all the angles in the smaller table?
   A 19°, 9.5°, and 61.5°
   B 38°, 19°, and 123°
   C 38°, 38°, and 104°
   D 76°, 38°, and 246°

5. Two rectangular gardens are similar. The area of the larger garden is 8.28 m², and its length is 6.9 m. The smaller garden is 0.6 m wide. What is the smaller garden's length and area?
   F length = 6.9 m; area = 2.07 m²
   G length = 3.45 m; area = 4.14 m²
   H length = 3.45 m; area = 1.97 m²
   J length = 3.45 m; area = 2.07 m²

6. Which of the following is not always true if two figures are similar?
   A They have the same shape.
   B They have the same size.
   C Their corresponding sides have proportional lengths.
   D Corresponding angles are congruent.

7. Which of the following figures are always similar?
   F two rectangles
   G two triangles
   H two squares
   J two pentagons

LESSON 7-4 Reading Strategies
Graphic Organizer

The information in this chart will help you understand **similar figures**.

Definition	Facts
Figures that have the same shape but may not be the same size	• Matching sides are proportional • Matching angles have the same measure.

Similar Figures

Example	Non-Example
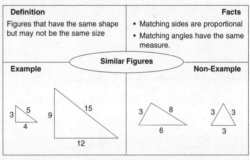	

Use the chart to help you answer these questions. Write *yes*, *no*, or *maybe*.

1. Do similar figures have the same shape?

   yes

2. Are similar figures the same size?

   maybe

3. Would a small square and a large square have different angle measurements?

   no

4. Would a small square and a large square have matching sides that are proportional?

   yes

5. Would a large square and a small square be examples of similar figures?

   yes

6. Would a large square and a large triangle be similar figures?

   no

LESSON 7-4 Puzzles, Twisters & Teasers
With One Blow!

Solve each of the problems below and circle your answer. Transfer the matching letters, in order, on to the blanks to solve the riddle.

1. A small triangle has a hypotenuse of 5, and sides of 3 and 4. A larger, similar, triangle has a hypotenuse of 30. Find the lengths of the other two sides of the larger triangle.
   K 25   (G) 24   F 12
   W 15   S 20   (I) 18

2. A large parallelogram has angles of 120 degrees and 60 degrees. What are the corresponding angles of a smaller, similar parallelogram?
   (A) 120   V 90   R 180
   P 30   E 360   (N) 60

3. Alok went to the photography store to develop some film. He has three choices of sizes for his prints. He thinks that two of the sizes make similar rectangles. Which size does *not* make a rectangle similar to the other two?
   M 4 by 6   (T) 8 by 10   B 8 by 12

What kind of ant can break a picnic table with one blow?

A G I A N T

LESSON 7-5 Practice A
Indirect Measurement

Write the correct answer.

1. Use similar triangles to find the height of the lamppost. $h = 10$ feet

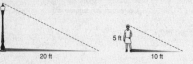

2. Use similar triangles to find the height of the man. $h = 6$ feet

3. A 3-foot-tall boy looks into a mirror at the county fair. The mirror makes a person appear shorter. The boy appears to be 1 foot tall in the mirror. If a man appears to be 2 feet tall in the mirror, what is his actual height?

   6 feet

4. On a sunny day, a carnation casts a shadow that is 20 inches long. At the same time, a 3-inch-tall tulip casts a shadow that is 12 inches long. How tall is the carnation?

   5 inches

5. A bicycle casts a shadow that is 8 feet long. At the same time, a girl who is 5 feet tall casts a shadow that is 10 feet long. How tall is the bicycle?

   4 feet

6. A sand castle casts a shadow that is 5 inches long. A 15-inch-tall bucket sitting next to the sand castle casts a shadow that is 3 inches long. How tall is the sand castle?

   25 inches

7. Through a magnifying glass, a 2-centimeter-long bug looks like it is 12 centimeters long. How long would a 3-centimeter bug look in that same magnifying glass?

   18 cm

8. In the late afternoon, a wagon casts a shadow that is 15 feet long. A boy pulling the wagon who is 4 feet tall casts a shadow that is 20 feet long. How tall is the wagon?

   3 feet

LESSON 7-5 Practice B
Indirect Measurement

Write the correct answer.

1. Use similar triangles to find the height of the building. $h = 24$ m

2. Use similar triangles to find the height of the taller tree.   **5 meters**

3. A lamppost casts a shadow that is 35 yards long. A 3-foot-tall mailbox casts a shadow that is 5 yards long. How tall is the lamppost?

   21 feet

4. A 6-foot-tall scarecrow in a farmer's field casts a shadow that is 21 feet long. A dog standing next to the scarecrow is 2 feet tall. How long is the dog's shadow?

   7 feet

5. A building casts a shadow that is 348 meters long. At the same time, a person who is 2 meters tall casts a shadow that is 6 meters long. How tall is the building?

   116 meters

6. On a sunny day, a tree casts a shadow that is 146 feet long. At the same time, a person who is 5.6 feet tall standing beside the tree casts a shadow that is 11.2 feet long. How tall is the tree?

   73 feet

7. In the early afternoon, a tree casts a shadow that is 2 feet long. A 4.2-foot-tall boy standing next to the tree casts a shadow that is 0.7 feet long. How tall is the tree?

   12 feet

8. Steve's pet parakeet is 100 mm tall. It casts a shadow that is 250 mm long. A cockatiel sitting next to the parakeet casts a shadow that is 450 mm long. How tall is the cockatiel?

   180 millimeters

LESSON 7-5 Practice C
Indirect Measurement

Write the correct answer.

1. Use similar triangles to find the height of the tower. $h = 29.76$ yd

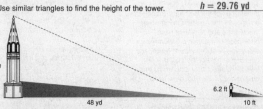

2. Use similar triangles to find the height of the man. $h = 5.4$ feet

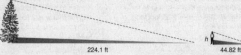

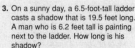

3. On a sunny day, a 6.5-foot-tall ladder casts a shadow that is 19.5 feet long. A man who is 6.2 feet tall is painting next to the ladder. How long is his shadow?

   18.6 feet

4. A building casts a shadow that is 1,125 meters long. A woman standing next to the building casts a shadow that is 6.25 meters long. She is 2.5 meters tall. How tall is the building?

   450 meters

5. Brian, who is twice as tall as Cole, is 6.5 feet tall. Cole casts a shadow that is 22.75 feet long. If Brian is standing next to Cole, how long is Brian's shadow?

   45.5 feet

6. A 4.5-foot-tall boy stands so the top of his shadow is even with the top of a flagpole's shadow. If the flagpole's shadow is 34 feet long, and the boy is standing 25 feet away from the flagpole, how tall is the flagpole?

   17 feet

7. A mother giraffe is 18.7 feet tall. Her baby is 5.25 feet tall. The baby giraffe casts a shadow that is 35.7 feet long. How long is the mother giraffe's shadow?

   127.16 feet

8. A shorter flagpole casts a shadow 15.3 feet shorter than the shadow of a longer pole. The taller pole is 26.5 feet tall and casts a shadow 47.7 feet long. How tall is the shorter pole?

   18 feet

LESSON 7-5 Reteach
Indirect Measurement

If you cannot measure a length directly, you can use indirect measurement. Indirect measurement uses similar figures and proportions to find lengths.

The small tree is 8 feet high and it casts a 12-foot shadow. The large tree casts a 36-foot shadow.

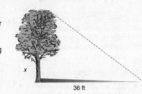

The triangles formed by the trees and the shadows are similar. So, their heights are proportional.

To find the height of the large tree, first set up a proportion. Use a variable to stand for the height of the large tree.

$\frac{8}{12} = \frac{x}{36}$ Write a proportion using corresponding sides.

$8 \cdot 36 = 12 \cdot x$ The cross products are equal.

$12x = 288$ x is multiplied by 12.

$\frac{12x}{12} = \frac{288}{12}$ Divide both sides by 12.

$x = 24$

So, the height of the tall tree is 24 feet.

Use indirect measurement to find the missing heights.

1.

150 feet

2.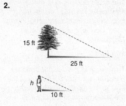

6 feet

LESSON 7-5 Challenge
Mirror Measurements

When it is noon, nighttime, a cloudy day, or when you are inside, there are hardly any shadows to use for indirect measurement. Instead, you can use mirrors to measure in the following way.

Place a mirror on the floor. Move back until you see the reflection of the top of the object you want to measure in the mirror. This creates two similar triangles. You can then use proportions to find the unknown height:

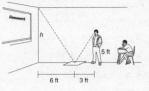

$\frac{h}{5} = \frac{6}{3}$

$h \cdot 3 = 5 \cdot 6$

$3h = 30$

$\frac{3h}{3} = \frac{30}{3}$

$h = 10$

So, the height of the classroom is 10 feet.

Find the missing height in each drawing to the nearest whole foot.

1.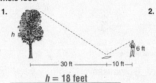

$h = 18$ feet

2.

$h = 12$ feet

3.

$h = 19$ feet

4.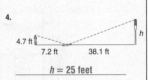

$h = 25$ feet

LESSON 7-5 Problem Solving
Indirect Measurement

Write the correct answer.

1. The Petronas Towers in Malaysia are the tallest buildings in the world. On a sunny day, the Petronas Towers cast shadows that are 4,428 feet long. A 6-foot-tall person standing by one building casts an 18-foot-long shadow. How tall are the Petronas Towers?

   1,476 feet

2. The Sears Tower in Chicago is the tallest building in the United States. On a sunny day, the Sears Tower casts a shadow that is 2,908 feet long. A 5-foot-tall person standing by the building casts a 10-foot-long shadow. How tall is the Sears Tower?

   1,454 feet

3. The world's tallest man cast a shadow that was 535 inches long. At the same time, a woman who was 5 feet 4 inches tall cast a shadow that was 320 inches long. How tall was the world's tallest man in feet and inches?

   8 feet 11 inches

4. Hoover Dam on the Colorado River casts a shadow that is 2,904 feet long. At the same time, an 18-foot-tall flagpole next to the dam casts a shadow that is 72 feet long. How tall is Hoover Dam?

   726 feet

Circle the letter of the correct answer.

5. An NFL goalpost casts a shadow that is 170 feet long. At the same time, a yardstick casts a shadow that is 51 feet long. How tall is an NFL goalpost?
   - A 100 feet
   - B 56 2/3 feet
   - **C 10 feet**
   - D 1 foot

6. A gorilla casts a shadow that is 600 centimeters long. A 92-centimeter-tall chimpanzee casts a shadow that is 276 centimeters long. What is the height of the gorilla in meters?
   - F 0.2 meter
   - **G 2 meters**
   - H 20 meters
   - J 200 meters

7. A 6-foot-tall man casts a shadow that is 30 feet long. If a boy standing next to the man casts a shadow that is 12 feet long, how tall is the boy?
   - A 2.2 feet
   - **C 2.4 feet**
   - B 5 feet
   - D 2 feet

8. An ostrich is 108 inches tall. If its shadow is 162 inches, and an emu standing next to it casts a 90-inch shadow, how tall is the emu?
   - F 162 inches
   - **H 60 inches**
   - G 90 inches
   - J 194.4 inches

LESSON 7-5 Reading Strategies
Following Procedures

Indirect means "not direct." We use **indirect measurement** when it is not possible to use standard measurement tools. Measuring the height of a very tall tree or water tower is difficult to do directly. An indirect method of measuring uses similar figures and proportions to find the length or height.

These two triangles are similar.

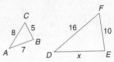

1. What side of the second triangle corresponds to side AB?   **side DE**

2. What side of the second triangle corresponds to side AC?   **side DF**

3. What side of the second triangle corresponds to side BC?   **side EF**

You can set up a proportion using two corresponding sides from each triangle to find the missing side.

4. Write a ratio using the lengths for sides AC and DF. $\frac{8}{16}$

5. Write a ratio using the lengths for sides AB and DE. $\frac{7}{x}$

6. Write the proportion for the above two ratios. $\frac{8}{16} = \frac{7}{x}$

7. Write a ratio using the lengths for sides BC and EF. $\frac{5}{10}$

8. Write a ratio using the lengths for sides AB and DE. $\frac{7}{x}$

9. Write a proportion for the above two ratios. $\frac{5}{10} = \frac{7}{x}$

LESSON 7-5 Puzzles, Twisters & Teasers
Not Quite the Same

Solve the crossword. Unscramble the circled letters to answer the question.

Down
1. In similar __FIGURES__ corresponding angles are congruent.
2. Figures that have the same shape but not necessarily the same size. __SIMILAR__
4. Solve: $\frac{5}{30} = \frac{x}{42}$ __SEVEN__
6. To sum two numbers. __ADD__
7. Six, fourteen, fifty-two, and seventy-eight are all __EVEN__ numbers.

Across
3. In similar figures, corresponding __SIDES__ have lengths that are proportional.
5. Indirect __MEASUREMENT__
8. Solve: $\frac{6}{x} = \frac{18}{33}$ __ELEVEN__
9. A figure with three sides is a __TRI__ -angle.

What do you use to make indirect measurements?

S I M I L A R FIGURES

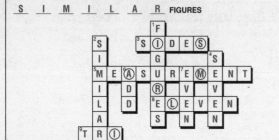

LESSON 7-6 Practice A
Scale Drawings and Maps

Use the map to answer the questions.

1. On the map, the distance from Newton to Cambridge is 2 cm. What is the actual distance?
   8 kilometers
2. On the map, the distance from Arlington to Medford is 1 cm. What is the actual distance?
   4 kilometers
3. If the distance between two cities on this map measures 6 centimeters, what is the actual distance?
   24 kilometers
4. If the actual distance between two cities is 12 kilometers, how many centimeters will separate those two cities on this map?
   3 centimeters

Use the scale drawing to answer each question.

5. This scale drawing is of the *Mayflower*, the ship that the first English settlers of Massachusetts used. How long was the actual *Mayflower*?
   100 feet
6. The height of the actual *Mayflower* was 200 feet from the bottom of the boat to the top of the tallest mast. Is the ship's height in the drawing correct?
   Yes; 2 in. = 200 ft

Scale: 1 inch = 100 feet

LESSON 7-6 Practice B
Scale Drawings and Maps

Use the map to answer the questions.

1. On the map, the distance between Big Cypress Swamp and Lake Okeechobee is $\frac{1}{4}$ inch. What is the actual distance?
   25 miles
2. On the map, the distance between Key West and Cuba is $\frac{9}{10}$ inch. What is the actual distance?
   90 miles
3. Use a ruler to measure the distance between Key West and Key Largo on the map. What is the actual distance?
   100 miles
4. The Overseas Highway connects Key West to mainland Florida. It is 110 miles long. If it were shown on this map, how many inches long would it be?
   $1\frac{1}{10}$ inches

Use the scale drawing to answer each question.

5. This scale drawing is of the lighthouse on Key West, originally built in 1825. What is the actual height of the lighthouse?
   85 feet
6. The original lighthouse was 66 feet tall. It was rebuilt at its present height after a hurricane destroyed it in 1846. How tall would the original lighthouse be in this scale drawing?
   $1\frac{13}{20}$ inches

1 inch = 40 feet

LESSON 7-6 Practice C
Scale Drawings and Maps

Use a metric ruler and the map to answer the questions.

1. What is the actual distance between Baltimore, Maryland, and Philadelphia, Pennsylvania?
   150 kilometers
2. What is the actual distance between Hartford, Connecticut, and Manchester, New Hampshire?
   160 kilometers
3. What is the actual distance between Washington, D.C., and Trenton, New Jersey?
   240 kilometers
4. Which actual distance is longer: New York City to Boston, or New York City to Washington, D.C.? What is the difference between those distances?
   New York City to Washington, D.C., is about 50 kilometers longer.

Use the scale drawing to answer each question.

5. What is the actual height of the Statue of Liberty, including its pedestal?
   300 feet
6. What is the actual height of the statue from the base to the tip of her torch?
   150 feet
7. The statue's arm that holds the torch is 42 feet long. How many inches long should it be on this drawing?
   0.42 in.

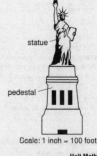
Scale: 1 inch = 100 foot

Holt Mathematics

LESSON 7-6 Reteach
Scale Drawings and Maps

A scale drawing is a drawing of a real object that is proportionally smaller or larger than the real object.

A scale is a ratio between two sets of measurements. In the map below, the scale is 2 cm: 0.5 km. This means that each centimeter on the map represents 0.25 kilometer.

Store

Park

School

Library

To find the actual distance from school to the library, first measure the distance on the map using a ruler.

The distance on the map is 3 centimeters.

$\frac{2 \text{ cm}}{0.5 \text{ km}} = \frac{3 \text{ cm}}{x \text{ km}}$ Write a proportion using the scale.

$2 \cdot x = 0.5 \cdot 3$ The cross products are equal.

$2x = 1.5$ x is multiplied by 2.

$\frac{2x}{2} = \frac{1.5}{2}$ Divide both sides by 2.

$x = 0.75$

The distance from school to the library is 0.75 kilometer.

Use the map to find each actual distance.

1. from the store to the library
   1 kilometer

2. from the park to the store
   1.25 kilometers

3. from the park to the library
   1.625 kilometers

4. from the park to the school
   2 kilometers

LESSON 7-6 Challenge
Solar System String

Distances in outer space are usually measured in millions of miles. Understanding or comparing such huge measurements can be difficult, and it is impossible to map or draw them in their actual scale. Here's an activity that can help you understand the vast scale of our solar system. Identify the 1-millimeter mark on your ruler. This tiny distance represents 1,000,000 miles in space! You will use it as the scale for your model: 1 millimeter = 1 million miles.

Make a scale model of our solar system.

1. Cut a piece of string 4 meters long. Tape a small piece of paper at one end of the string and label it "Sun."
2. From the sun, measure 3.6 cm. Tape a "Mercury" label there.
3. From Mercury, measure another 3.1 cm. Tape a "Venus" label there.
4. From Venus, measure another 2.6 cm. Tape an "Earth" label there.
5. From Earth, measure another 4.9 mm. Tape a "Mars" label there.
6. From Mars, measure another 34.2 cm. Tape a "Jupiter" label there.
7. From Jupiter, measure another 40.2 cm. Tape a "Saturn" label there.
8. From Saturn, measure another 89.8 cm. Tape a "Uranus" label there.
9. From Uranus, measure another 1,010 mm. Tape a "Neptune" label there.
10. From Neptune, measure another 88.1 cm. Tape a "Pluto" label there.

Now use the scale and your model to find the actual distance from Earth. For example, the distance from Earth to the sun on the string measures 93 mm, so the actual distance is 93 million miles.

Earth to Mercury: **57 million miles**
Earth to Venus: **26 million miles**
Earth to Mars: **4.9 million miles**
Earth to Jupiter: **346.9 million miles**
Earth to Saturn: **748.9 million miles**
Earth to Uranus: **1,646.9 million miles**
Earth to Neptune: **2,656.9 million miles**
Earth to Pluto: **3,537.9 million miles**

LESSON 7-6 Problem Solving
Scale Drawings and Maps

Write the correct answer.

1. About how many kilometers long is the northern border of California along Oregon?
   about 300 kilometers

2. What is the distance in kilometers from Los Angeles to San Francisco?
   about 500 kilometers

3. How many kilometers would you have to drive to get from San Diego to Sacramento?
   about 750 kilometers

4. At its longest point, about how many kilometers long is Death Valley National Park?
   about 200 kilometers

5. Approximately what is the distance, in kilometers, between Redwood National Park and Yosemite National Park?
   about 350 kilometers

Circle the letter of the correct answer.

6. Which of the following two cities in California are about 200 kilometers apart?
   (A) San Diego and Los Angeles
   B Monterey and Los Angeles
   C San Francisco and Fresno
   D Palm Springs and Bakersfield

7. Joshua Tree National Park is about 200 kilometers from Sequoia National Park. How many centimeters should separate those parks on this map?
   F 110 cm
   G 11 cm
   (H) 1 cm
   J 0.11 cm

LESSON 7-6 Reading Strategies
Use Graphic Aids

A **scale drawing** is larger or smaller than the actual object. The shape of the drawing is the same as the actual object. The scale determines the size of the drawing.

This map is an example of a scale drawing. Each centimeter on the map stands for 10 kilometers. The map scale ratio is $\frac{1 \text{ cm}}{10 \text{ km}}$.

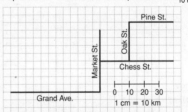

Answer each question to set up a proportion and find out how many kilometers long Market Street is.

1. What is the map scale ratio? $\frac{1 \text{ cm}}{10 \text{ km}}$

2. Measure Market Street with a centimeter ruler. How many centimeters long is it?   **4 cm**

3. Make a ratio with the map length of Market Street on the top of the ratio and the length in kilometers of Market Street (x) on the bottom. $\frac{4 \text{ cm}}{x}$

4. Use the ratios from Exercises 1 and 3 to write a proportion. $\frac{1 \text{ cm}}{10 \text{ km}} = \frac{4 \text{ cm}}{x \text{ km}}$

Answer each question to find out how many kilometers long Grand Avenue is.

5. How many centimeters long is Grand Avenue on the map?   **6 cm**

6. Write a proportion using the map scale ratio and x divided by the map measurement of Grand Avenue. $\frac{10}{1} = \frac{x}{6}$

LESSON 7-6 Puzzles, Twisters & Teasers
In Order, Please!

On each grid below, find the actual length of the lines. (The grids have different scales.) Place the *actual* lengths in order from smallest to largest. Use the corresponding letters to solve the riddle.

Grid #1-scale: 1 unit = 5 feet

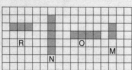

R = 15, N = 25, O = 20, M = 10

Grid #2-scale: 1 unit = 7 feet

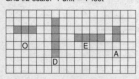

O = 21, D = 35, E = 28, A = 14

What happened when a ship carrying a load of blue paint collided with a ship carrying a load of red paint?

The crews were M A R O O N E D .

LESSON 7-7 Practice A
Percents

Use the 10-by-10-square grids to model each percent.

1. 12%
2. 67%

Write each percent as a fraction in simplest form.

3. 50% $\frac{1}{2}$
4. 1% $\frac{1}{100}$
5. 11% $\frac{11}{100}$

6. 10% $\frac{1}{10}$
7. 99% $\frac{99}{100}$
8. 17% $\frac{17}{100}$

Write each percent as a decimal.

9. 5%   0.05
10. 75%   0.75
11. 2%   0.02

12. 15%   0.15
13. 13%   0.13
14. 90%   0.90

15. The math workbook has 100 pages. Each chapter of the book is 10 pages long. What percent of the book does each chapter make up?

    10%

16. There were 100 questions on the math test. Chi-Tang answered 88 of those questions correctly. What percent did he get correct on the test?

    88%

LESSON 7-7 Practice B
Percents

Write each percent as a fraction in simplest form.

1. 30% $\frac{3}{10}$
2. 42% $\frac{21}{50}$
3. 18% $\frac{9}{50}$

4. 35% $\frac{7}{20}$
5. 100% $\frac{1}{1}$ or 1
6. 29% $\frac{29}{100}$

7. 56% $\frac{14}{25}$
8. 70% $\frac{7}{10}$
9. 25% $\frac{1}{4}$

Write each percent as a decimal.

10. 19%   0.19
11. 45%   0.45
12. 3%   0.03

13. 80%   0.8
14. 24%   0.24
15. 6%   0.06

Order the percents from least to greatest.

16. 89%, 42%, 91%, 27%

    27%, 42%, 89%, 91%

17. 2%, 55%, 63%, 31%

    2%, 31%, 55%, 63%

18. Sarah correctly answered 84% of the questions on her math test. What fraction of the test questions did she answer correctly? Write your answer in simplest form.

    $\frac{21}{25}$

19. Chloe swam 40 laps in the pool, but this was only 50% of her total swimming workout. How many more laps does she still need to swim?

    40 more laps

LESSON 7-7 Practice C
Percents

Write each percent as a fraction or mixed number in simplest form.

1. 68% $\frac{17}{25}$
2. 98% $\frac{49}{50}$
3. 55% $\frac{11}{20}$

4. 84% $\frac{21}{25}$
5. 16% $\frac{4}{25}$
6. 60% $\frac{3}{5}$

7. 125% $1\frac{1}{4}$
8. 150% $1\frac{1}{2}$
9. 140% $1\frac{2}{5}$

Write each percent as a decimal.

10. 0.5%   0.005
11. 0.25% 0.0025
12. 127%   1.27

13. 205%   2.05
14. 1165% 11.65
15. 0.08% 0.0008

Order from least to greatest.

16. 92%, 0.86, 47%, and $\frac{14}{25}$

    47%, $\frac{14}{25}$, 0.86, 92%

17. 5%, $\frac{7}{100}$, 0.8%, 0.003

    0.003, 0.8%, 5%, $\frac{7}{100}$

18. Of all the students who voted for their favorite ice cream, $\frac{1}{2}$ chose chocolate and $\frac{2}{5}$ chose vanilla. What percent of all the votes were not for chocolate or vanilla?

    10% of the votes

19. On his last 5 math quizzes, Paulo got the following scores: 97%, $\frac{3}{4}$, 82%, $\frac{91}{100}$, and $\frac{7}{10}$. What was his average quiz score?

    83%

Reteach 7-7 Percents

A percent is a ratio of a number to 100. Percent means "per hundred."

To write 38% as a fraction, write a fraction with a denominator of 100.
$$\frac{38}{100}$$
Then write the fraction in simplest form.
$$\frac{38}{100} = \frac{38 \div 2}{100 \div 2} = \frac{19}{50}$$
So, $38\% = \frac{19}{50}$.

Write each percent as a fraction in simplest form.

1. 43% $\frac{43}{100}$
2. 72% $\frac{18}{25}$
3. 88% $\frac{22}{25}$
4. 35% $\frac{7}{20}$

To write 38% as a decimal, first write it as fraction.
$$38\% = \frac{38}{100}$$
$\frac{38}{100}$ means "38 divided by 100."

```
   0.38
100)38.00
   -300
    800
   -800
      0
```
So, 38% = 0.38.

Write each percent as a decimal.

5. 64% 0.64
6. 92% 0.92
7. 73% 0.73
8. 33% 0.33

Challenge 7-7 Per State

To show a percent, you can shade a 10-by-10 grid in any design that you want. For each percent below, try to shade the grid to look like the state it describes.

1. California has the largest population of any state. About 12% of all Americans live in California.

Possible 12% shading for California

2. Florida is the top tourist state. About 26% of all visitors to the United States choose Florida for their vacations.

Possible 26% shading for Florida

3. Nevada is the fastest-growing state. Its population has grown about 66% in the last ten years.

Possible 66% shading for Nevada

4. Alaska is the largest state. It makes up about 15% of the total area of the United States.

Possible 15% shading for Alaska

5. Washington produces the most apples. About 50% of all the apples grown in the U.S. come from Washington.

Possible 50% shading for Washington

6. Texas is the top oil-producing state. About 21% of all the oil produced in the United States comes from Texas.

Possible 21% shading for Texas

Problem Solving 7-7 Percents

Use the circle graph to answer each question. Write fractions in simplest form.

1. What fraction of the total 2000 music sales in the United States were rock recordings?
   $\frac{1}{4}$

2. On this grid, model the percent of total United States music sales that were rap recordings. Then write that percent as a decimal.
   0.13

U.S. Recorded Music Sales, 2000
Oldie 1%, Classical 3%, Other 18%, Rock 25%, Jazz 3%, Rap 13%, Religious 5%, R&B 10%, Pop 11%, Country 11%

Circle the letter of the correct answer.

3. What kind of music made up $\frac{1}{20}$ of the total U.S. music recording sales?
   A Oldie
   B Classical
   C Jazz
   D Religious

4. What fraction of the United States music sales were country recordings?
   F $\frac{110}{100}$
   G $\frac{11}{100}$
   H $\frac{1}{10}$
   J $\frac{1}{100}$

5. What fraction of all United States recording sales did jazz and classical music make up together?
   A $\frac{6}{10}$
   B $\frac{3}{50}$
   C $\frac{1}{5}$
   D $\frac{11}{100}$

6. What kind of music made up $\frac{1}{10}$ of the total music recording sales in the United States in 2000?
   F Pop
   G Jazz
   H R&B
   J Oldies

Reading Strategies 7-7 Use Graphic Aids

The word **percent** means "per hundred." It is a ratio that compares a number to 100. A grid with 100 squares is used to picture percents.

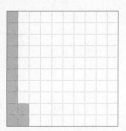
Twelve percent is pictured on the grid below.

12 percent is a ratio, and means 12 per hundred. → $\frac{12}{100}$

12 percent can be written with symbols. → 12%

Use this figure to complete Exercises 1–4.

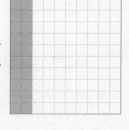

1. What is the ratio of shaded squares to the total number of squares? $\frac{20}{100}$

2. Write the shaded amount using the % symbol.
   20%

3. What is the ratio of unshaded squares to total number of squares? $\frac{80}{100}$

4. Use the % symbol to write the unshaded amount. 80%

LESSON 7-7 Puzzles, Twisters & Teasers
Perfect 100!

Decide whether each statement is true or false. Circle your answer.

Unscramble your circled letters to spell an important word to know.

1. $45\% < \frac{1}{2}$ (N) true   B false
2. $6.4 = 64\%$ A true   (R) false
3. A 7% tax rate means you will pay $7 tax on a $10 purchase.   W true   (T) false
4. $\frac{1}{5} = 20\%$ (P) true   H false
5. Thirty-seven percent is greater than one-third.   (E) true   I false
6. Three-quarters is more than 80%.   K true   (C) false
7. $22\% = \frac{2}{11}$ M true   (E) false

Answer: __PERCENT__

LESSON 7-8 Practice A
Percents, Decimals, and Fractions

Write each decimal as a percent.

1. 0.1 __10%__
2. 0.6 __60%__
3. 0.02 __2%__
4. 0.14 __14%__
5. 0.22 __22%__
6. 0.03 __3%__
7. 0.25 __25%__
8. 0.17 __17%__
9. 0.39 __39%__
10. 0.8 __80%__
11. 0.04 __4%__
12. 0.99 __99%__

Write each fraction as a percent.

13. $\frac{1}{2}$ __50%__
14. $\frac{1}{4}$ __25%__
15. $\frac{3}{4}$ __75%__
16. $\frac{7}{10}$ __70%__
17. $\frac{97}{100}$ __97%__
18. $\frac{33}{100}$ __33%__

19. Brett scored $\frac{1}{4}$ of all the baskets he shot during the basketball game. What percent did he make?
__25%__

20. Sarah has 3 dimes and 1 nickel. Jamie has 2 quarters. What percent of a dollar do they each have?
__Sarah has 35% of a dollar, and Jamie has 50% of a dollar.__

21. Mike, Joey, and Kini are playing a shooting game at the fair. Mike made $\frac{3}{5}$ of his shots, Joey made $\frac{4}{5}$, and Kini made $\frac{2}{5}$. Write the percent each boy made.
__Mike 60%, Joey 80%, Kini 40%__

LESSON 7-8 Practice B
Percents, Decimals, and Fractions

Write each decimal as a percent.

1. 0.03 __3%__
2. 0.92 __92%__
3. 0.18 __18%__
4. 0.49 __49%__
5. 0.7 __70%__
6. 0.09 __9%__
7. 0.26 __26%__
8. 0.11 __11%__
9. 1.0 __100%__

Write each fraction as a percent.

10. $\frac{2}{5}$ __40%__
11. $\frac{1}{5}$ __20%__
12. $\frac{7}{10}$ __70%__
13. $\frac{1}{20}$ __5%__
14. $\frac{1}{50}$ __2%__
15. $\frac{4}{50}$ __8%__

Compare. Write <, >, or =.

16. 60% __<__ $\frac{2}{3}$
17. 0.4 __=__ $\frac{2}{5}$
18. 0.5 __>__ 5%
19. $\frac{1}{100}$ __<__ 0.03
20. $\frac{7}{9}$ __>__ 72%
21. $\frac{3}{10}$ __<__ 35%

22. Bradley completed $\frac{3}{5}$ of his homework. What percent of his homework does he still need to complete?
__40%__

23. After reading a book for English class, 100 students were asked whether or not they enjoyed it. Nine twenty-fifths of the students did not like the book. How many students liked the book?
__64 students__

LESSON 7-8 Practice C
Percents, Decimals, and Fractions

Write each decimal as a percent and as a fraction or mixed number.

1. 0.96 __96%; $\frac{24}{25}$__
2. 0.04 __4%; $\frac{1}{25}$__
3. 0.28 __28%; $\frac{7}{25}$__
4. 0.65 __65%; $\frac{13}{20}$__
5. 0.32 __32%; $\frac{8}{25}$__
6. 0.005 __0.5%; $\frac{1}{200}$__
7. 1.13 __113%; $1\frac{13}{100}$__
8. 2.08 __208%; $2\frac{2}{25}$__
9. 3.002 __300.2%; $3\frac{1}{500}$__

Write each fraction as a percent and as a decimal. Round to the nearest hundredth if necessary.

10. $\frac{12}{13}$ __92%; 0.92__
11. $\frac{22}{27}$ __81%; 0.81__
12. $\frac{15}{26}$ __58%; 0.58__
13. $\frac{9}{31}$ __29%; 0.29__
14. $\frac{34}{35}$ __97%; 0.97__
15. $\frac{11}{23}$ __48%; 0.48__

Compare. Write <, >, or =.

16. $\frac{12}{17}$ __<__ 77%
17. 0.18 __<__ $\frac{11}{25}$
18. $\frac{11}{50}$ __=__ 0.22
19. $\frac{21}{33}$ __<__ 80%
20. 0.4 __>__ $\frac{9}{32}$
21. $\frac{5}{16}$ __>__ 28%

22. During a sale, everything in the store was $\frac{1}{5}$ off the ticketed price. What percent of an item's original price should you expect to pay?
__80%__

23. Your teacher has offered you a choice for your 50 homework problems. You can do 48% of the problems, all of the even-numbered problems, or $\frac{3}{5}$ of the problems. Which option will you choose? How many problems will you have to do for homework?
__Answers will vary, but most students will choose the option with the fewest problems: 48% of the problems or 24 problems.__

Reteach
7-8 Percents, Decimals, and Fractions

You can write decimals as percents.
To write 0.5 as a percent, multiply the decimal by 100%.
$0.5 \cdot 100\% = 50\%$
To multiply a number by 100, move the decimal point two places to the right.
0.50
So, 0.5 = 50%.

Write each decimal as a percent.

1. 0.8 — 80%
2. 0.64 — 64%
3. 0.075 — 7.5%
4. 0.29 — 29%

You can solve a proportion to write a fraction as a percent.
To write $\frac{3}{4}$ as a percent, first set up a proportion.

$\frac{3}{4} = \frac{x}{100}$
$3 \cdot 100 = 4 \cdot x$ — The cross products are equal.
$300 = 4x$ — x is multiplied by 4.
$\frac{4x}{4} = \frac{300}{4}$ — Divide both sides by 4.
$x = 75$
So, $\frac{3}{4} = \frac{75}{100}$
$\frac{75}{100} = 75\%$, So, $\frac{3}{4} = 75\%$.

Write each fraction as a percent.

5. $\frac{4}{5}$ — 80%
6. $\frac{9}{10}$ — 90%
7. $\frac{1}{8}$ — 12.5%
8. $\frac{7}{25}$ — 28%
9. $\frac{1}{4}$ — 25%
10. $\frac{5}{6}$ — 83.3%
11. $\frac{3}{4}$ — 75%
12. $\frac{1}{5}$ — 20%

Challenge
7-8 Trash or Treasure?

People in the United States produce about 208 million tons of garbage every year! We recycle about 56 million tons of that garbage, or about 27% of the total.

Complete the chart at right. Then display the percents on the circle graph below. Remember to give your graph a title. Label each section of the graph with the material and the percent of the total garbage recycled that each section represents. You may wish to color each section differently or add illustrations.

United States Recycling

Material	Total Garbage Recycled	
	Fraction	Percent
Metals	$\frac{1}{10}$	10%
Yard Waste	$\frac{17}{100}$	17%
Glass	$\frac{3}{50}$	6%
Paper	$\frac{29}{50}$	58%
Plastics	$\frac{1}{50}$	2%
All Other Materials	$\frac{7}{100}$	7%

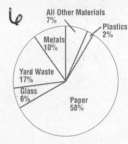

United States Recycling
- All Other Materials 7%
- Plastics 2%
- Metals 10%
- Yard Waste 17%
- Glass 6%
- Paper 58%

Problem Solving
7-8 Percents, Decimals, and Fractions

Write the correct answer.

1. Deserts cover about $\frac{1}{7}$ of all the land on Earth. About what percent of Earth's land is made up of deserts?
   about 14%

2. The Sahara is the largest desert in the world. It covers about 3% of the total area of Africa. What decimal expresses this percent?
   0.03

3. Cactus plants survive in deserts by storing water in their thick stems. In fact, water makes up $\frac{3}{4}$ of the saguaro cactus's total weight. What percent of its weight is water?
   75%

4. Daytime temperatures in the Sahara can reach 130°F! At night, however, the temperature can drop by 62%. What decimal expresses this percent?
   0.62

Circle the letter of the correct answer.

5. The desert nation of Saudi Arabia is the world's largest oil producer. About $\frac{1}{4}$ of all the oil imported to the United States is shipped from Saudi Arabia. What percent of our nation's oil is that?
   A 20%
   B 22%
   C 25%
   D 40%

6. About $\frac{2}{5}$ of all the food produced on Earth is grown on irrigated cropland. What percent of the world's food production relies on irrigation? What is the percent written as a decimal?
   F 40%; 40.0
   G 40%; 4.0
   H 40%; 0.4
   J 40%; 0.04

7. About $\frac{3}{25}$ of all the freshwater in the United States is used for drinking, washing, and other domestic purposes. What percent of our fresh water resources is that?
   A 3%
   B 25%
   C 12%
   D $\frac{1}{5}$

8. Factories and other industrial users account for about $\frac{23}{50}$ of the total water usage in the United States. Which of the following show that amount as a percent and decimal?
   F 46% and 0.46
   G 23% and 0.23
   H 50% and 0.5
   J 46% and 4.6

Reading Strategies
7-8 Multiple Meanings

A person can go by different names. Timothy could also be called Tim or Timmy.

A number can have different names too. The columns below show different names for 0.4. → Read: "four tenths."

Decimal Form	Fraction Form	Percent Form
0.4 →	$\frac{4}{10}$ →	40%

To find the percent form of a fraction or decimal number, write an equivalent fraction with a denominator of 100.

$\frac{4}{10} = \frac{40}{100}$

A fraction with a denominator of 100 can be written as a percent. → $\frac{4}{10}$ → 40%

Use 0.37 to complete Exercises 1–3.

1. Write the words for 0.37. — **thirty-seven hundredths**
2. Write 0.37 as a fraction. — **$\frac{37}{100}$**
3. Write 0.37 as a percent. — **37%**

Use $\frac{60}{100}$ to complete Exercises 4–7.

4. How would you read $\frac{60}{100}$? — **sixty hundredths**
5. Write $\frac{60}{100}$ as a decimal. — **0.60**
6. Write $\frac{60}{100}$ as a percent. — **60%**

---

99 Holt Mathematics

Lesson 7-8 Puzzles, Twisters & Teasers
Chitter-Chatter!

Some people talk all the time, while others never say a word. Have you ever wondered which animal talks the most? You are about to find out!

In the box below, cross out any number that is greater than one. 121%, 2.06, $\frac{3}{2}$

Next, cross out any pairs that have equivalent values. [$\frac{1}{4}$, 25%] [$\frac{3}{4}$, 75%]
(For example, $\frac{1}{2}$ is the same as 50%).

Finally, order the remaining numbers from smallest to largest and place the corresponding letters in the answer spaces below. [0.16, $\frac{1}{3}$, 53%, 0.88]

```
    ¾ D            ⅓ Y
         121% P              53% A
  0.88 K           ¼ T
              75% V
                        0.16 A
                0.25 E
    2.06% M            3/2 R
```

Place the remaining numbers (you should have 4 left) in order from smallest to largest in the spaces below.

What animal talks the most?

__A__ __Y__ __A__ __K__

Lesson 7-9 Practice A
Percent Problems

Find the percent of each number.
1. 10% of 30 __3__
2. 30% of 90 __27__
3. 20% of 40 __8__
4. 50% of 14 __7__
5. 2% of 10 __0.2__
6. 15% of 6 __0.9__
7. 5% of 20 __1__
8. 60% of 10 __6__
9. 50% of 50 __25__
10. 4% of 4 __0.16__
11. 90% of 10 __9__
12. 10% of 25 __2.5__
13. 25% of 100 __25__
14. 70% of 10 __7__
15. 75% of 100 __75__
16. 35% of 15 __5.25__
17. 25% of 20 __5__
18. 8% of 16 __1.28__

19. Courtney made 12 model racecars. She painted 75% of her cars blue. How many of Courtney's racecar models are blue?
__9 cars are blue__

20. Tim used 16 large beads to make a necklace. He chose bright orange for 25% of those beads. How many beads on Tim's necklace are bright orange?
__4 beads are bright orange__

21. Taylor has 25 stuffed animals. She took 20% of those animals with her to a slumber party. How many stuffed animals did Taylor take to the slumber party?
__5 stuffed animals__

Lesson 7-9 Practice B
Percent Problems

Find the percent of each number.
1. 8% of 40 __3.2__
2. 105% of 80 __84__
3. 35% of 300 __105__
4. 13% of 66 __8.58__
5. 64% of 50 __32__
6. 51% of 445 __226.95__
7. 14% of 56 __7.84__
8. 98% of 72 __70.56__
9. 24% of 230 __55.2__
10. 35% of 225 __78.75__
11. 44% of 89 __39.16__
12. 3% of 114 __3.42__
13. 70% of 68 __47.6__
14. 1.5% of 300 __4.5__
15. 85% of 240 __204__
16. 47% of 13 __6.11__
17. 20% of 522 __104.4__
18. 2.5% of 400 __10__

19. Jenna ordered 28 shirts for her soccer team. Seventy-five percent of those shirts were size large. How many large shirts did Jenna order?
__21 large shirts__

20. Douglas sold 125 sandwiches to raise money for his boy scout troop. Eighty percent of those sandwiches were sold in his neighborhood. How many sandwiches did Douglas sell in his neighborhood?
__100 sandwiches__

21. Samuel has run for 45 minutes. If he has completed 60% of his run, how many minutes will Samuel run in all?
__75 minutes__

Lesson 7-9 Practice C
Percent Problems

Find the percent of each number.
1. 22% of 22 __4.84__
2. 147% of 600 __882__
3. 16% of 48 __7.68__
4. 65% of 1,185 __770.25__
5. 96% of 12 __11.52__
6. 9% of 29 __2.61__
7. 25% of 455 __113.75__
8. 77% of 326 __251.02__
9. 87% of 113 __98.31__
10. 15.6% of 470 __73.32__
11. 92% of 514 __472.88__
12. 2.5% of 16 __0.4__
13. 7.2% of 65 __4.68__
14. 84.2% of 65 __54.73__
15. 4.5% of 880 __39.6__
16. 36.8% of 400 __147.2__
17. 6.5% of 250 __16.25__
18. 211% of 22 __46.42__

19. The soccer team ordered 140 T-shirts to sell at the school fair. Of those T-shirts, 50% are white, 20% are blue, 15% are green, 10% are red, and 5% are black. How many black T-shirts did the soccer team order? how many red?
__7 black T-shirts and 14 red T-shirts__

20. The Johnsons ordered new carpet for their family room. They paid 33% of the total cost when they ordered it, and will pay the remaining amount when the carpet is delivered. If they paid $192.72 when they ordered the carpet, how much will they pay when it is delivered?
__$391.28__

21. The city is going to raise its sales tax from 6.25% to 8.5% after the first of the year. How much tax would someone save on a $19,540 car if they bought the car before the first of the year rather than after the first of the year?
__$439.65__

Reteach
7-9 Percent Problems

You can use proportions to solve percent problems.
To find 25% of 72, first set up a proportion.

$\frac{25}{100} = \frac{x}{72}$

$25 \cdot 72 = 100 \cdot x$ Next, find cross products.

$1,800 = 100x$

$\frac{100x}{100} = \frac{1,800}{100}$ Then solve the equation.

$x = 18$

So, 18 is 25% of 72.

Use a proportion to find each number.

1. Find 3% of 75. 2. Find 15% of 85. 3. Find 20% of 50. 4. Find 6% of 90.

   2.25   12.75   10   5.4

You can use multiplication to solve percent problems.
To find 9% of 70, first write the percent as a decimal.
$9\% = 0.09$
Then multiply using the decimal.
$0.09 \cdot 70 = 6.3$
So, 9% of 70 = 6.3.

Use multiplication to find each number.

5. Find 80% of 48. 6. Find 6% of 30. 7. Find 40% of 120. 8. Find 20% of 98.

   38.4   1.8   48   19.6

9. Find 70% of 70. 10. Find 35% of 120. 11. Find 9% of 50. 12. Find 40% of 150.

   49   42   4.5   60

Challenge
7-9 Pet Percentages

The United States Census Bureau counts all the people in the United States—but they do not count our pets! So, veterinarians use the percents shown in the table below to estimate pet populations. Their estimated U.S. pet population data is based on the 2000 census, which counted about 106 million households in the United States.

U.S. Pet Census, 2000

Pet	Percent of all Households	Estimated U.S. Pet Population
Dogs	53%	56,180,000
Cats	60%	63,600,000
Birds	13%	13,780,000
Horses	4%	4,240,000

Use the percents to estimate the number of pets that your class owns altogether, and the number of pets that your school owns altogether. Let each student in your class and each student in your school represent 1 household.

My Class and School Pet Population   use 200

Pet	Estimated Class Pet Population	Estimated School Pet Population
Dogs	15.9	106
Cats	18	120
Birds	3.9	26
Horses	1.2	8

Answers will vary depending on the number of students in the class and the number of students in the school. Example answers are given for a 30-student class and a 200-student school.

Problem Solving
7-9 Percent Problems

In 2000, the population of the United States was about 280 million people.
Use this information to answer each question.

1. About 20% of the total United States population is 14 years old or younger. How many people is that?

   56 million people

2. About 6% of the total United States population is 75 years old or older. How many people is that?

   16.8 million people

3. About 50% of Americans live in states that border the Atlantic or Pacific Ocean. How many people is that?

   140 million people

4. About 12% of all Americans live in California. What is the population of California?

   33.6 million people

5. About 7.5% of all Americans live in the New York City metropolitan area. What is the population of that region?

   21 million people

6. About 12.3% of all Americans have Hispanic ancestors. What is the Hispanic American population here?

   34.44 million people

Circle the letter of the correct answer.

7. Males make up about 49% of the total population of the United States. How many males live here?
   A 1,372 million   C 13.72 million
   (B) 137.2 million   D 1.372 million

8. About 75% of all Americans live in urban areas. How many Americans live in or near large cities?
   F 70 milliom   (H) 210 million
   G 200 million   J 420 million

9. About 7.4% of all Americans live in Texas. What is the population of Texas?
   A 74 million   C 7.4 million
   (B) 20.72 million   D 2.072 million

10. Between 1990 and 2000, the population of the United States grew by about 12%. What was the U.S. population in 1990?
    (F) 250 million   H 313.6 million
    G 33.6 million   J 268 million

Reading Strategies
7-9 Connect Words and Symbols

There are 24 students in Mrs. Wilson's class. Twenty-five percent of them take the bus to school. How many students ride the bus to school?

Step 1: Write a statement for the problem. → 25% of 24 students take the bus.

Step 2: Write an equation for the problem. → 25% of 24 is what number?

Step 3: Use symbols in place of words. → $25\% \cdot 24 = x$

Step 4: Change the percent to a decimal. → $0.25 \cdot 24 = x$

Step 5: Multiply to solve. → $x = 6$

Answer each question.

1. What symbol stands for "of"?

   $\cdot$

2. What does x stand for in this problem?

   the number of students who ride the bus

3. Write the decimal value for 25%.

   0.25

30% of the class brings their lunch to school. There are 50 sixth graders in the class. How many students bring their lunch?

4. Write a statement for this problem.

   30% of 50 is what number?

5. Rewrite the statement, using symbols for "of" and "is".

   $30\% \cdot 50 = x$

6. Rewrite the problem using a decimal in place of 30%.

   $0.30 \cdot 50 = x$

7. Multiply to solve. How many students bring their lunch?

   15 students

Lesson 7-9 Puzzles, Twisters & Teasers: Odd Man Out!

In the problems below, two of the expressions mean the same thing, but one is different. Circle the one that is different. Write the circled letters in the corresponding spaces to solve the riddle.

1. 20% of 500 pages **A**   0.2 · 500 pages **W**   1,000 pages **(P)**
2. $\frac{40}{1.25}$ **(L)**   50 **D**   125% of 40 **T**
3. 20 minutes **R**   30% of an hour **(N)** $\frac{1}{3}$ of an hour **F**
4. 40% complete **S**   job is half-done **(V)**   60% more to do **I**
5. 20 ÷ 5 discount **(O)**   20% discount **K** $\frac{1}{5}$ reduction in price **M**
6. $\frac{3.6}{9}$ **B**   40% **H**   4 **(E)**

What starts with "E", ends with "E," but contains only one letter?

E N V E L O P E E
_ 3 4 6 2 5 1 _

Lesson 7-10 Practice A: Using Percents

Write the correct answer.

1. Glenn bought some candy that cost $5.00. If he had to pay a 5% sales tax, how much did he pay for his candy in all?
   $5.25

2. Nathan has ordered a pizza that costs $20.00. He wants to give the delivery person a 20% tip. How much should the tip be?
   $4.00

3. Jasmine wanted to buy a new pair of shoes that cost $30.00. When she went to the store, she found that the shoes were on sale for 10% off. How much did Jasmin pay for her shoes?
   $27.00

4. Marie ordered a root beer float at the ice cream shop. The float was $2.00, and she paid a 6% sales tax. How much did Marie pay for her root beer float in all?
   $2.12

5. Taylor has a coupon for 15% off any item in the toy store. The remote-control airplane he wants is $40.00. How much will the airplane cost if Taylor uses his coupon?
   $34.00

6. Victor went to the barber to get a haircut. The haircut cost $9.00. Victor gave the barber a 25% tip. How much did Victor spend at the barber shop altogether?
   $11.25

7. A CD is on sale for $10.00. The sales tax rate is 6%. How much will the total cost be for the CD?
   $10.60

8. A video game costs $25.00. The sales tax is 7%. How much will the total cost be for the game?
   $26.75

9. A bead store has a sign that reads "10% off the regular price." If Janice wants to buy beads that regularly cost $6.00, how much will she pay for them after the store's discount?
   $5.40

10. Julie gets a 20% discount on all of the items in the clothing store where she works. If she buys a shirt that regularly costs $45.00, how much money will she save with her employee discount?
    $9.00

Lesson 7-10 Practice B: Using Percents

Write the correct answer.

1. Carl and Rita ate breakfast at the local diner. Their bill came to $11.48. They gave their waitress a tip that was 25% of the bill. How much money did they give the waitress for her tip?
   $2.87

2. The school's goal for the charity fundraiser was $3,000. They exceeded the goal by 22%. How much money for charity did the school raise at the event?
   $3,660

3. Rob had a 15% off coupon for the sporting goods store. He bought a tennis racket that had a regular ticket price of $94.00. How much did Rob spend on the racket after using his coupon?
   $79.90

4. Lisa's family ordered sandwiches to be delivered. The total bill was $21.85. They gave the delivery person a tip that was 20% of the bill. How much did they tip the delivery person?
   $4.37

5. A portable CD player costs $118.26. The sales tax rate is 7%. About how much will it cost to buy the CD player?
   $126.54

6. Kathy bought two CDs that each cost $14.95. The sales tax rate was 5%. About how much did Kathy pay in all?
   $31.40

7. Tom bought $65.86 worth of books at the book fair. He got a 12% discount since he volunteered at the fair. About how much did Tom's books cost after the discount?
   $57.96

8. Sawyer bought a T-shirt for $12.78 and shorts for $17.97. The sales tax rate was 6%. About how much money did Sawyer spend altogether?
   $32.60

9. Melody buys a skateboard that costs $79.81 and a helmet that costs $26.41. She uses a 45% off coupon on the purchase. If Melody pays with a $100 bill, about how much change should she get back?
   $41.58

10. Bruce saved $35.00 to buy a new video game. The game's original price was $42.00, but it was on sale for 30% off. The sales tax rate was 5%. Did Bruce have enough money to buy the game? Explain.
    Yes; with the discount and sales tax, the total cost was $30.87.

Lesson 7-10 Practice C: Using Percents

Write the correct answer. **Estimates may vary.**

1. A computer costs $979.99. The sales tax rate is 7%. How much will the total cost be for the computer?
   $1,050

2. Sheila bought $146.87 worth of groceries. The sales tax rate was 6%. How much did she spend in all?
   $155

3. Paul has saved $37.50 to buy a hamster, a cage, and hamster food. The hamster cost $5.50. The cage is on sale for 25% off the original price of $29.90. The food cost $2.64. The sales tax on the total is 5%. How much will Paul pay in all? How much of his savings will he have left over?
   $32.09; $5.41

4. Jake has $65.50 to buy a new pair of jeans and a shirt. The jeans he wants cost $42.50, and the shirt costs $29.50. He has a coupon for 15% off, and the sales tax is 5%. Will he have enough money? Explain.
   Yes, the total cost is $64.26, which is less than Jake has.

5. The bike Henry wants usually costs $147.99. Today, it is on sale for 15% off. After an 8% sales tax, how much will Henry pay for the bike?
   $135.85

6. Scott's lunch bill is $11.79. He gets an employee discount of 10% off. He leaves a 20% tip for the waitress. How much does Scott spend for lunch in all?
   $12.73

7. At Paint City, a gallon of paint with a regular price of $17.99 is now 15% off. At Giant Hardware, the same paint usually costs $21.99, but is now 24% off. Which store is offering the better deal?
   Paint City

8. Chelsea and Raymond's dinner bill was $57.82. They left the waitress a 26% tip. If they split the total cost of dinner evenly, how much did they each pay?
   $36.43

9. A store is having a going out of business sale for 55% off the ticketed prices. A pair of in-line skates has a ticketed price of $59.85, and a scooter has a ticketed price of $64.80. With a sales tax of 5%, how much will it cost to buy both items?
   $58.90

10. Troy works in a sporting goods store and gets an employee discount of 15% off any purchase. He wants to buy a fishing pole with a regular price of $70.00. The fishing pole is on sale for 33% off. How much will Troy pay for the fishing pole?
    $39.87

---

102 Holt Mathematics

Reteach
7-10 Using Percents

There are many uses for percents.

Common Uses of Percents

Discounts	A **discount** is an amount that is subtracted from the regular price of an item. discount = regular price • discount rate
Tips	A **tip** is an amount added to a bill. tip = total bill • tip rate
Sales Tax	**Sales tax** is an amount added to the price of an item. sales tax = purchase price • sales tax rate

Rachel is buying a sweater that costs $42. The sales tax rate is 5%. About how much will the total cost of the sweater be?

You can use fractions to find the amount of sales tax.

First round $42 to $40.

Think: 5% is equal to $\frac{1}{20}$.

So, the amount of tax is about $\frac{1}{20}$ • $40.

The tax is about $2.00.

Then find the sum of the price of the sweater and the tax.

$42 + $2.00 = $44.00

Rachel will pay about $44.00 for the sweater.

Solve each problem. Estimates may vary.

1. About how much would you pay for a meal that costs $29.75 if you left a 15% tip?

   $34.50

2. About how much do you save if a book whose regular price is $25.00 is on sale for 10% off?

   $2.50

3. About how much would you pay for a box of markers whose price is $5.99 with a sales tax rate of 9.5%?

   $6.60

Challenge
7-10 Shop Smart

The Sport Zone and Sport City are both competing for customers by offering big discounts. To be a smart customer, you need to decide which store is offering the better price on each item.

For each item, write the store offering the best deal and the price you will pay there to the nearest whole cent.

The Sport Zone   Sport City

1.    The Sport Zone; $118.99

2.    Sport City; $30.30

3.   The Sport Zone; $15.50

4.   Sport City; $42.20

Problem Solving
7-10 Using Percents

Use the table to answer each question.

Federal Income Tax Rates, 2001

Single Income	Tax Rate	Married Joint Income	Tax Rate
$0 to $27,050	15%	$0 to $45,200	15%
$27,051 to $65,550	27.5%	$45,201 to $109,250	27.5%
$65,551 to $136,740	30.5%	$109,251 to $166,500	30.5%
$136,741 to $297,350	35.5%	$166,501 to $297,350	35.5%
More than $297,350	39.1%	More than $297,350	31.5%

1. If a single person makes $25,000 a year, how much federal income tax will he or she have to pay?

   $3,750

2. If a married couple makes $148,000 together, how much federal income tax will they have to pay?

   $45,140

3. The average salary for a public school teacher in the United States is $42,898. If two teachers are married, what is the average amount of federal income taxes they have to pay together?

   $23,593.90

4. In 2002 President George W. Bush received an annual salary of $400,000. Vice President Dick Cheney got $186,300. How much federal income tax do they each have to pay on their salary if they are married and filing jointly?

   Bush: $126,000;
   Cheney: $66,136.50

Circle the letter of the correct answer.

5. Members of the U.S. Congress each earn $145,100 a year. How much federal income tax does each pay on their salary?
   A $51,510.50   C $21,765
   B $44,255.50   D $39,902.50

6. A married couple each working a minimum-wage job will earn an average of $21,424 together a year. How much income tax will they pay?
   F $5,891.60   H $321.36
   G $3,213.60   J $6,534.32

7. The average American with a college degree earns $33,365 a year. About how much federal income tax does he or she have to pay at a single rate?
   A $5,004.75   C $10,176.33
   B $9,175.38   D $11,844.58

8. The governor of New York makes $179,000 a year. How much federal income tax does that governor have to pay at a single rate?
   F $63,545   H $49,225
   G $54,595   J $26,850

Reading Strategies
7-10 Use a Graphic Organizer

This chart shows common ways percents are used. It also shows you how to figure a discount, sales price, sales tax, total price, tip, and total cost of a meal.

Discount
An amount subtracted from the regular price of an item
- Discount = regular price • discount rate
- Sales price = regular price − the discount

Sales Tax
An amount added to the price of an item
- Sales tax = purchase price • sales tax rate
- Total price = regular price + sales tax

Uses for Percents

Tip
The amount added to a bill for service
- Tip = price of meal • tip rate

Use the graphic organizer to answer each question.

1. What do you call the amount subtracted from the regular price of an item?

   discount

2. What do you call the amount added to a bill for service?

   tip

3. What do you call an amount added to the price of an item?

   sales tax

4. How do you find the discount?

   multiply the regular price times the discount rate

5. How can you find the amount of sales tax on an item?

   multiply the purchase price times the sales tax rate

6. How is a total bill for a meal figured?

   add the price of the meal and the tip

LESSON 7-10 **Puzzles, Twisters & Teasers**

Teacher's Favorite!

Decide whether each statement is true or false. If the statement is true, follow the directions to navigate the maze. If the statement is false, ignore the directions and go to the next problem. Unscramble the letters that you land on to solve the riddle.

1. A tip is an amount added to a bill. Begin at start and move four spaces up. **True**

2. A discount is an amount added to a bill. Move five spaces diagonally down and to the left. **False**

3. Sihla is buying several CDs, totaling $45. If the sales tax is 5%, she will pay $47.25 total. Move three spaces left. Then move three spaces diagonally down and to the left. **True**

4. You can find 10% of a number by moving the decimal point one place to the right. Move 6 spaces up. **False**

5. A sign in a store reads "15% off all items". This is the same as a 15 percent discount on all items. Move three spaces diagonally down and to the right. Then move to the right as far as you can. **True**

What is your teacher's favorite candy?

C H A L K --OLATE